Raise

*The publisher gratefully acknowledges
the generous support of the General
Endowment Fund of the University of
California Press Foundation.*

Raise

*What 4-H Teaches Seven Million Kids
and How Its Lessons Could Change
Food and Farming Forever*

KIERA BUTLER

With photographs by Rafael Roy

UNIVERSITY OF CALIFORNIA PRESS

University of California Press, one of the most
distinguished university presses in the United
States, enriches lives around the world by
advancing scholarship in the humanities, social
sciences, and natural sciences. Its activities are
supported by the UC Press Foundation and by
philanthropic contributions from individuals
and institutions. For more information, visit
www.ucpress.edu.

University of California Press
Oakland, California

Library of Congress Cataloging-in-Publication Data

Butler, Kiera, 1980–
Raise : what 4-H teaches seven million kids and
how its lessons could change food and farming
forever / Kiera Butler ; with photographs by
Rafael Roy.
   pages   cm
Includes bibliographical references.
ISBN 978-0-520-27580-5 (pbk : alk.
paper) — ISBN 978-0-520-95898-2 (e-book)
1. 4-H clubs—United States.   2. Agricultural
education—United States.   I. Roy, Rafael.   II. Title.
S533.F66B88   2014
630.71'1—dc23   2014010237

Manufactured in the United States of America

23 22 21 20 19 18 17 16 15 14
10 9 8 7 6 5 4 3 2 1

The paper used in this publication meets the
minimum requirements of ANSI/NISO Z39.48-1992
(R 2002) (*Permanence of Paper*).

For Meta and JB, and for Mom and Dad

CONTENTS

This book is the result of the two-plus years I spent pestering 4-H'ers and their families: I followed the kids around as they did their chores, asked each of them hundreds of questions over bowls of ice cream and cups of hot chocolate, and distinguished myself at the fairs as the only ringside specta-tor with an open laptop. To recreate the scenes and conversations in this book, I drew on a mix of my recordings, videos, photos, and interview notes. For context and supplemental material, I conducted dozens of interviews with 4-H program leaders, livestock experts, academics, and others—many of whom I quote in the text. To the best of my knowledge, everything I present here is factually accurate. The book has been independently fact checked.

In a few cases, I've changed the names of minor characters to respect their privacy. Although much of the book's action proceeds in the order in which it actually happened, there are a few exceptions where I sped up cer-tain periods of time and condensed others for the sake of clarity. For exam-ple: I met Randy Sosa of Greenfield 4-H later in the course of my research than he turns up in the book, and I did not go to Ghana until after I was done with most of my other reporting.

Parts of this book—the section about my turkeys and the few paragraphs about the Thode family of Sebastopol, California—emerged from blog posts I first wrote for *Mother Jones* magazine.

I am not a historian, and this book isn't a conventional history. Much of the historical material I've included is from Marilyn and Thomas Wessel's 1982 book, *4-H: An American Idea, 1900–1980*.

Introduction

One sunny day in May, two roommates and I drove out to a farm in the country to pick up a couple of three-week-old turkey poults. The farmer raised her eyebrows when we told her that we planned to raise them in our urban neighborhood in Berkeley, California. But we assured her that our turkeys would have a good life, and she relinquished the apple-sized birds to us. They chirped plaintively throughout the entire hour and a half of our drive home.

But it didn't take long for our turkeys to get used to their new life in Berkeley. Until they were big enough to live outside, we kept them in our living room. There is a reason that people don't typically keep poultry in their living rooms: the birds poop everywhere and make your house smell awful. But what our turkeys lacked in personal hygiene they more than made up for with charm. In the evenings, they followed my two roommates and me from room to room, skittering along in an adorable half-flying, half-scurrying fashion. We taught them to fly from one couch to another. If I curled up in a chair to read a book, the birds would eventually arrange themselves on my lap and fall asleep. They looked like miniature dinosaurs.

When they were ready for an outdoor home, our across-the-street neighbors, a carpenter and a landscaper, built an impressive and spacious pen for the turkeys in their yard. I came over every morning with treats for the birds: raisins, nuts, and canned tuna for extra protein. They developed a particular taste for Trader Joe's arugula, which they could spot before I even took it out of the bag. They flapped their wings and nipped at me in anticipation. Both birds turned out to be hens—lucky for us and our neighbors, since male birds would have made a lot more noise.

Despite their bohemian upbringing, they thrived. By November 10 they were fifteen pounds each, big enough to feed a crowd. As Thanksgiving approached, we all agreed that we wanted to do right by these amazing creatures.

When we pictured killing our turkeys, we envisioned a somber and beautiful ceremony honoring the lives of the two birds we had raised since they were small enough to fit in the palms of our hands. We had researched the most humane way to end their lives—by hanging them upside down in a cone and slitting their jugular vein with a single fast and merciful cut. I had picked out a luminous W. S. Merwin poem for the occasion. There was talk of burning sage.

But on the afternoon when we assembled to slaughter the first of the two turkeys, something happened that we hadn't accounted for: kids started showing up. Word had spread in our neighborhood that something was going to be killed, and everyone wanted to be there to watch the spectacle.

Bikes were ditched in front of our neighbors' house. John, a seven-year-old I had met a few times, wandered into the yard. He'd heard that the killing was imminent, and he had a lot of questions.

"Did they ever bite you?" he wanted to know. "Did you bleed? Is that why you're killing them?"

"When are you gonna do it?" asked a kid on a scooter.

"Oooh, don't let him bite me, I'm gonna get him!" squealed another, hustling away from the turkey, who had found some onion greens to nibble on. A few kids horsed around by the vegetable bed. There was a lot of yelling.

"What do we do?" I hissed at my friend. The problem was not just that the kids were spoiling our plans for a solemn ceremony. We also weren't sure about the ethical propriety of letting a bunch of kids witness us killing an animal. We considered asking the kids to get their parents' permission. But tracking them down could have taken a long time. With some ambivalence, we decided to go ahead with the plan.

One little girl of about twelve shyly sidled up to me as I was trying to keep the turkey calm. There was something touching about watching the kids take it all in. Morbid curiosity was surely part of the draw ("This is gonna be sick!"). But there was reverence, too.

"Those feathers are pretty," said the girl. "What do they feel like?"

"They're soft," I said. "You want to touch them?"

She gingerly reached out and patted the turkey. "They *are* soft."

We watched as my carpenter neighbor mounted a traffic cone onto an old spiraling iron railing—he had found it at a junkyard—and placed a metal pot below to catch the blood. He caught the bird and lowered her into the cone headfirst. I didn't observe the actual slitting of her throat; I was busy whisking the other turkey away so she wouldn't see. (Understandably, birds can become stressed out by witnessing slaughter.)

By the time I got back, blood was draining out of the cone and into the pot. John, the curious seven-year-old, came up close to have a look.

"What if the cops come?" he said.

"This isn't illegal, duh!" said an older kid, rolling his eyes.

"Well, what if it was a person?"

"You are so stupid. Why you gotta be so dumb?" Lots of giggles.

"Oooh, I can see chunks in there," said John, peering into the pot of blood. "That is so nasty!" I marveled at how none of the kids seemed squeamish. They didn't shrink away from the goriest of details.

Once the bird had bled out, the kids took off on their scooters and bikes.

When we killed the other turkey a few days later, it was just us grownups. I read my W. S. Merwin poem. Everyone was quiet and respectful. But I found myself missing the kids' curiosity and energy.

Our Thanksgiving celebration was incredible. The upstairs neighbor, a chef, wrapped one of the turkeys in a layer of bacon, followed by a layer of cheesecloth, and finally a layer of peanut butter. He roasted the whole thing. We were skeptical—especially of the peanut butter part. But when he took the bird out of the oven, everyone agreed it was the most delicious Thanksgiving turkey they'd ever had. I like to attribute this success to the months of careful nurturing and frequent arugula treats rather than the preparation, but in truth it was probably a combination of the two. Our across-the-street neighbors ate the other turkey for their Thanksgiving (sans peanut butter), and they said it turned out delicious as well. I wished the neighborhood kids could have been there to share the meal with us.

A few weeks after Thanksgiving, I ran into John on my way home from work.

"You gonna get another turkey?" he asked.

"I don't know," I said. "Maybe next year."

He seemed satisfied with this answer and rode off on his bike.

What was it about the turkey that had captured John's attention? Was it the grisly death? Or was it that he had made a connection between our bird and what his family would likely eat for Thanksgiving dinner a few days later?

I hoped it had been the latter. As I would later learn, fewer and fewer American kids have the chance to see where their food comes from. And the effects of this disconnect are not pretty.

The year is 2012, the scene a bustling community center day camp in a major US city. The camp has a special visitor today, a researcher from the University of California, Davis, who has set up two chairs in a corner of a classroom.[1] She has been conducting one-on-one interviews with the students all morning.

The researcher introduces herself to an eleven-year-old named Lilly, who slips into the chair facing her. She asks Lilly if she's ever seen a cow.

Lilly nods, and the researcher smiles encouragingly. She asks Lilly whether she's seen different kinds of cows, and Lilly says she has.

"What was different about them?" she asks.

"Well," Lilly says carefully, "one of them had brown spots instead of black, but you know how cows are all white and have little black spots? Some of them have brown spots."

The researcher nods. "So is there a difference?"

"Yeah," says Lilly. "You know how some have a kind of pink look? I think that's where they get that kind of milk. And they get, I think they get chocolate milk from the one that has brown spots."

"Okay," says the researcher. "So which one would you pick for your farm?"

"I would pick the one that got brown spots because the one that has black spots just gives out regular white milk, but the one with brown spots gives out chocolate milk," says Lilly.

"So cows have different colors and the different colors indicate different milk flavors, and you would pick cows that have the flavor of milk you like for your farm?" asks the researcher. She wants to make sure she has understood Lilly correctly.

"Yeah," says Lilly.

"Why do you think farmers chose the plants and animals they raise on their farms?" asks the researcher.

"The same, I guess," says Lilly. "They like the taste of the colors."

The researcher thanks Lilly and sends her back to her camp group. The counselor sends over the next camper, a ten-year-old named Art. The researcher smiles brightly and asks, "Why do you think farmers select the animals and plants that they grow?"

Art hesitates, trying to figure out what this researcher expects of him. "So they can make money when they sell them to the factories," he finally decides.

"Okay," says the researcher. "Is there any other reason why farmers may decide to select the plants or animals that they grow?"

Art thinks for a minute, and then his face lights up. "Maybe because the sheep have fur so the factory can make pillows."

The interviews continue like this all afternoon. At the end of the day, the researcher has spoken to eighteen campers between the ages of nine and eleven. A few weeks later, she and her colleagues write up their findings. They compare the children's knowledge to benchmarks for their age group set by the American Association for the Advancement of Sciences and a body of agricultural researchers. They note that overall, the interviewees "held no discernable understanding that crops came from different parts of the world, had biologic origins, and are often derived from different cultural groups." They conclude that the interviewees' basic understanding of the role of science and technology in agriculture is "nearly non-existent."

It is disheartening to learn that an eleven-year-old believes that cows with brown spots produce chocolate milk, but it could be that these particular children were especially uninformed about agriculture. After all, they were city kids, like John and his friends who had watched us slaughter our turkeys. The researchers had determined that their farm experience had been limited mostly to school field trips.

Unfortunately, these children are not an exception. By now, it is well known that Americans young and old are a little fuzzy on our food's origins. But the extent of the disconnect between people and farms is truly shocking. It is estimated that 90 percent of Americans come from families that haven't farmed for two or three generations.[2] A 2011 survey by the US Farmers and Ranchers Alliance (USFRA) found that 72 percent of Americans say they know nothing or very little about agriculture.[3] In a 2005 poll, the W. K. Kellogg Foundation found that "Americans think very little about where their food comes from."[4] Consider these two examples from the interviews the foundation's researchers conducted with adults:

Q: When you think about the food that you eat, what are the steps that get it to your table, whether it's bread or produce or whatever?

A: Oh, you mean like the steps—in other words shopping, or microwave?

Q: More like, How was it produced? How did it end up in the grocery store?

A: That, I'm not as familiar with.

Urban female, age 33

A: Fish? Where does it come from? Well, it comes from the ocean.

Q: Sure. But so how does it get here?

A: Oh um, I never really thought of that. I guess they fish for it.

Q: OK. Who do you picture fishing?

A: I don't know, kind of I guess just fishermen. I don't know exactly.

Q: What's the picture in your head if there is one?

A: Well, when I do buy fish, I normally buy it at a Chinese market. It's like an indoor market. It's a supermarket or like a grocery store. But it's just for Chinese food . . . and if I'm going to buy shrimp or fish or anything I'll buy it there. Because it's fresh.

Urban female, age 37

Some of the respondents showed wishful thinking about their food. For example:

Q: Would you have any sense of where the milk that you would buy would come from?

A: Local dairy, local cows I would guess.

Q: If you picture where those cows are what comes to mind?

A: They're in a nice green pasture somewhere.

Q: Do you figure that's probably the truth or is that how you would like to picture it?

A: That's the way I'd like to picture it. They probably eat out of a trough and don't wander around very much.

Urban male, age 62

And:

Q: Do you feel like the food sources are safe?

A: For the most part. You hear reports of people getting sick from South American strawberries and whatnot. I buy all those foods, and I've never really gotten sick from any foods. So for the most part, by my personal experience, I feel the food supply is pretty safe.

Urban male, age 34

In 2010, Texas State University administered a basic agricultural literacy test to 501 members of its incoming freshman class.[5] The average score for the section about food, nutrition, and health was just 40 percent. The most hopeful sign that the study's authors could wring from the results was that "the relatively high score of 55.7% for Theme 1 'Understanding Agriculture' indicated that there might be a general understanding that agriculture plays a role in everyday life among the respondents."

When researchers asked New York State high school students to describe a typical farmer in a 2009 survey, the most common responses "involved a man wearing overalls, a straw hat and a plaid shirt, with hay sticking out of his mouth and a pitchfork in his hand. Tractors and big red barns were also mentioned. The words *redneck*, *hick*, and *hillbilly* were used in many cases as well."[6] Studies have found that teachers harbor similar stereotypes. In 2010, researchers from Oregon State University asked teachers to describe what they thought a farm looked like.[7] One teacher said, "When I think of a farm, I think of a big red barn." Another said that all she knew about farms was "what I've seen on TV, *Little House on the Prairie* like."

It's easy to bemoan the state of American ignorance about farming. But how do we change it? A good group to start with would be children, who are excellent learners. It would be helpful if public schools taught students about food systems, but in these days of dwindling education budgets, it's unlikely that agricultural literacy will become a priority anytime soon. Some high schools and a small number of middle schools have chapters of the National FFA Organization (formerly Future Farmers of America), a youth club that teaches agriculture and animal husbandry through hands-on projects. But FFA chapters are mostly located in farming communities.

The answer might lie outside schools, in one organization with both a history of agriculture education and unparalleled access to American youth: 4-H. Founded at the turn of the nineteenth century, 4-H began as a network of agriculture clubs for the children of farmers. The earliest 4-H members competed in corn-growing and food-canning competitions, applying the latest advances in farming from the universities to their own lives. Over the past century, the club has evolved to keep up with the shifting lifestyles and interests of American youth.

Today, through a unique public-private structure, 4-H is officially housed under the National Institute of Food and Agriculture (NIFA) within the US Department of Agriculture (USDA), and it administers its programs largely through the federal Cooperative Extension programs at the nation's land-grant universities—the 111 schools that receive federal grant money to teach agriculture, science, and engineering. With 6.5 million members in all fifty states and another half million participants in more than seventy countries, 4-H is one of the largest youth development organizations, if not the largest.[8] (By comparison, Girl Scouts has 2.3 million US members; Boy Scouts has 2.7 million.) The group boasts 3,500 staff and 538,000 volunteers. If living 4-H alumni had their own country, it would be the size of Italy. The extensive list of notable 4-H alumni includes many present and former members of Congress, secretaries of agriculture, and state politicians.[9] Other famous former 4-H'ers are Dolly Parton, Julia Roberts, Jacqueline Kennedy Onassis, *Garfield* creator Jim Davis, John Updike, and Al Gore.

Farm kids no longer make up the bulk of 4-H membership: Fewer than half of 4-H members live on farms and in rural areas. Just shy of a quarter come from a city with a population of more than fifty thousand, and the remainder live in smaller cities and suburbs.[10] And not all of 4-H's programs have to do with agriculture; 4-H'ers can make a project out of virtually any topic that interests them and their club volunteers and leaders. Web design, calligraphy, GPS navigation, scrapbooking, and shooting sports are just a few of the projects offered by clubs that I found in California.[11] But farm education was the club's original raison d'être, and it remains a major focus. Even 4-H'ers who compete in the organization's junk-drawer robotics competition or enroll in their club's model rocket project learn about agriculture; 4-H leaders have figured out that science education and agricultural literacy go hand in hand. The organization is deeply involved not only with the science of growing crops and raising animals but also with the business of farming.

The National 4-H Council, the nonprofit that supports most of 4-H's major nationwide initiatives, gets the majority of its money from its corporate sponsors—among which are virtually every big name in agribusiness: Monsanto, DuPont, John Deere, the United Soybean Board, Cargill, Philip Morris, and Walmart, to name just a few. In return for their sponsorship, 4-H offers these businesses valuable services.[12] As national 4-H program leader Jim Kahler put it to me, "Monsanto and DuPont have a vested interest in keeping kids interested in science, since agriculture is science today. You're doing robotics, so we try to make the connection between that robot you're building and a robot on a farm." The firms' perspective and input shape what 4-H'ers learn about science and agriculture: they hear a lot about the benefits of industrial farming and biotechnology—and little about the environmental and social consequences.

When I set out to write this book, I intended to tell a story about kids raising animals. In part, that's what I did. But, as I discovered, there was another important story: how one of America's oldest agricultural organizations is teaching agriculture to America's children—children like John, the little boy who was fascinated by my turkeys, and Lilly, the girl who thought that chocolate milk came from cows with brown spots. As I learned more about 4-H, I began to wonder how the organization was shaping kids' understanding of farms.

In the grand scheme of Lilly's education, making sure she understands chocolate milk might seem like a relatively unimportant goal. But in less than a decade, Lilly will go to the supermarket, where her decisions will affect not only her own health and that of her family but also the status quo of the American food system and its cascade of global impacts. Farms cover 922 million acres of land in the United States—that's about a third of the country, an area nearly the size of India. The United States is the largest exporter of agricultural products in the world; it supplies a third of the world's corn and half of its soybeans.[13] About 15 percent of American workers are employed in the food and fiber industries.

Lilly's purchases will influence manufacturers' choices about which chemicals to put into foods, what standards of safety to adhere to, and how to raise the livestock that becomes meat. Lilly will vote, and her ballot will help shape legislation that determines the amount of food aid for needy Americans, the distribution of government subsidies for farmers, and the rules that determine how we take care of one of our most valuable natural resources: the American land. Lilly might even choose one of a growing number of agricultural jobs, in which case she herself could revolutionize the vast systems that produce our food, clothes, and a host of other things we use every day. Lilly's agricultural education doesn't concern just Lilly—in fact, it will influence all of us.

The story I tell in this book focuses mostly on 4-H in California—with a brief detour to Ghana, where 4-H has recently expanded its presence—although the nation's 4-H'ers are as diverse as the hundreds of thousands of towns they come from. I don't pretend to have written a definitive work on 4-H; many of the group's pockets were beyond my scope of reporting. The group's robust clubs in the agricultural heartland of the Midwest, for example, deserve their own book. But California's 4-H community is vast and varied—its members are from urban, suburban, and agricultural neighborhoods. The source of a third of the produce grown in the United States, California is certainly a farming region worth paying attention to.

So what are the world's seven million 4-H'ers—kids like Lilly—learning about farms? I set out asking that question. This book is the result of what I found.

"I Wanted to Be a Cowgirl"

A confession: I reached the age of thirty-one without ever having darkened the doorstep of a county fair. I was living in Berkeley, California, where we had street festivals, which I studiously avoided on account of the annoying naked people who always showed up. Where I was raised, in the unleafy city of Somerville, Massachusetts, we had yearly carnivals with rides and maybe a depressing pony or two. Fairs, I had long believed, were a treat accessible only to people who lived in the country.

So when I happened to see a sign for the Alameda County Fair, just a half-hour drive east of Berkeley, in the suburb of Pleasanton, I made up my mind to go. One hot July day, two friends and I piled into my car bound for the fair.

Once inside the fairgrounds, I realized that this was no mere traveling carnival. Hordes of people streamed past us toward a cavernous building that promised exhibits having to do with "food and fiber arts." Another building was dedicated to minerals, gems, and history. Still another was labeled "Model Railroad." In an open space, kids in harnesses dangled on bungee cords over an enormous trampoline. Hot tub salesmen showed off their jets.

We debated our next move—games? rides? funnel cake?—when out of the corner of my eye I noticed a large, open-air structure labeled Amador Livestock Pavilion. I pointed to the sign excitedly and began striding toward it. My indulgent friends trailed behind.

You might be wondering why a grown woman was displaying toddler-like delight at the prospect of seeing barnyard animals. The truth is that I was going through a livestock phase. It was 2011. It seemed like everyone I knew was talking about farming. Michael Pollan's influential book *The Omnivore's Dilemma* had come out five years earlier, and more recently, journalist and farmer Novella Carpenter had written her book *Farm City* about raising turkeys, goats, and even pigs in one of Oakland's grittiest neighborhoods.

Chicken coops sprouted like toadstools overnight in my neighbors' yards. Hipsters began haunting the farmers markets, flirting with each other over baskets of kale and collard greens. The poor butcher at the organic meat stall couldn't get a moment's peace, so constantly surrounded was he by swooning girls. My friends and I signed up for weekly boxes of farm produce, which resulted in acrobatic feats of menu planning. For a while, if I typed "too much" into Google on my laptop, it automatically filled in "kohlrabi."

My friends and I built a chicken coop from scratch and raised eight hens and, later, our two turkeys. We went to classes at a local urban farming supply store. We gave one another homesteading manuals as birthday presents, and we devoured *Farm City*.

So you can see why, like a lamb to a field of alfalfa, I was drawn to the livestock pavilion. Outside, we passed a few teenage girls soaping up sheep on what looked like small wooden platforms. Inside, we wandered up and down the lines of neat pens. Eventually we reached "Champion Row," where a few pigs, goats, and sheep milled about their beribboned enclosures. I looked at these champions, then back at the others, trying to pick out the winning quality that distinguished these specimens from the leagues of loser animals behind them. Were they fatter? Cuter? I couldn't tell.

Poster boards outside the pens announced that each row of animals was affiliated with a 4-H club. I passed by a line of pigs that belonged to the

Mountain House 4-H club in Byron, California, then one from the Palomares 4-H club in Castro Valley.

I had only a very dim idea of what 4-H actually was. I knew it had something to do with raising animals, but I assumed that this took place in the country towns of states like Iowa, not in the Bay Area suburbs.

In one of the pens, a teenage girl dressed in a white uniform with a green hat and tie was fiddling with her sheep's water bucket. I asked if all the animals belonged to 4-H clubs.

"Yeah, pretty much," she said, giving me a teenage why-are-you-talking-to-me look. She went back to the water bowl. I continued down the row. When I turned the corner, I came face-to-face with a line of smaller girls, also in white uniforms, leading burly goats on leashes. The girls couldn't have been much older than ten or eleven; the goats seemed twice their size, yet the girls seemed to be in total control. When one of the animals strained and looked ready to bolt, its pigtailed owner gave a firm yank and it fell back into step.

I followed the signs pointing toward the small-animals area, which looked like a miniature version of the livestock pavilion. Instead of pens, there were rows of cages holding rabbits, guinea pigs, and chickens. I headed for the turkey area and found a science-fair-style three-paneled poster all about how to ensure that your turkey gets enough protein. At the top was a photo of a little boy in a white uniform holding a baby turkey, and toward the bottom were pictures of the grown-up bird, with lustrous white feathers and alert eyes.

It was as though I had entered a parallel farming universe. Right here in the suburbs was a network of kids quietly raising animals and producing literature about their protein needs.

And what was I? A city girl with a flock of impulse poultry. I was feeding them *canned tuna*, for God's sake. These kids—in their tidy white uniforms with their no-nonsense sheep platforms—were the real deal. And yet, not once in any of my urban farming classes, homesteading books, or flirtatious chats with farmers' market vendors had anyone mentioned 4-H kids.

I made up my mind to meet them.

Fortuitously, one Saturday afternoon a few weeks later, I happened to run into my neighbor Kristen, who lived on my street with her husband, Matt, and her eight-year-old son, Caton. I told her about my visit to the fair.

"Oh," she said. "We were there, too. For 4-H."

"4-H!" I gasped. "I know them!"

Kristen told me that this was Caton's first year as a member of the 4-H club in Montclair, a neighborhood in the Oakland hills. He was in the club's rocketry project and outdoor adventures group, which Matt was in charge of. She hoped that when Caton was nine, 4-H's minimum age for livestock projects, he might join the club's goat group.

"So would you keep the goats, like, in your yard?" I asked hopefully. I would have loved to have some neighborhood goats.

"We've thought about it," she said. "But we wouldn't have to." She explained that there was a central place where kids in Oakland kept their goats, up in the hills in some old lady's yard.

A 4-H club in Oakland! I asked Kristen if she could put me in touch with the leaders. In the meantime, I got my hands on a copy of a 1982 book called *4-H: An American Idea: 1900–1980* and began to learn about the club's history.[1]

To me, the single most surprising thing about 4-H's origin story is that in the beginning, the club was not about kids. It involved kids—but they weren't the main point. The kids were simply the delivery method for the club's larger mission.

The 4-H story begins at the turn of the last century, when researchers at universities throughout the heartland were beginning to experiment with the idea of using science to determine the best ways to farm. They analyzed soil composition, created test plots to identify the highest-yielding crop varieties, and pressed newly invented machinery into service during harvesting. The new methods, the researchers hoped, would increase farms' efficiency and lighten workloads.[2]

The nation's farmers, however, had been working the land in the same way for generations, their methods imported by their forebears from Europe. They resented university outsiders peddling machines. The researchers found that even the few farmers who were willing to embrace the new ways were hard to reach geographically, scattered at the ends of winding dirt roads or perched on the edge of the forest.

In 1898, Will B. Otwell, the president of the farmers' institute in Macoupin County, Illinois, and a proponent of the new farming methods, was having trouble convincing farmers to come to his meetings. After months of sparse attendance, Otwell had an idea: he would invite the farmers' children instead. He offered a prize of one dollar for the child who could grow the best corn from his seed. According to Marilyn and Thomas Wessel, the authors of the 4-H history book, the contest was a great success. Over the course of the following decade, Otwell brought the concept to neighboring states. In 1905, he sponsored a national competition, and tens of thousands of children from eight states entered their corn.[3]

Meanwhile, in the rural township of Springfield, Ohio, a superintendent of schools named Albert B. Graham noticed in 1902 that, despite the fact that most of his students came from farm families, agriculture was not part of the curriculum in any local school. He began to offer classes on the weekends to teach his students modern farming methods. Eventually, administrators at an agriculture experiment station at Ohio State University got wind of the program. Soon more than a dozen counties in Ohio had similar classes. Farm experiment stations in other states began to hear about Graham's work and started offering classes of their own.[4]

Other superintendents of rural school districts liked the idea because enrollments at schoolhouses were beginning to decline. In his 1906 book, *Among Country Schools*, O. J. Kern, the superintendent in Illinois's Winnebago County, raised the questions that were on administrators' minds.[5] "Has the improvement of the country school kept pace with other things?" he asked. "If so, why are so many people leaving the farm and moving to the cities to educate their children?"

Kern believed that the reputation of farming was suffering and that unless the profession could be brought into the modern era, more and more farmers would throw down their hoes in disgust and take the next train into the city.[6] Kern's most notable accomplishment was the Winnebago Farmer Boys' Experiment Club, an extracurricular group that he organized in 1902 to "try to interest the big boys in the work of the Illinois College of Agriculture and the Experiment Station." Acting as a liaison between the college and the local country schools, Kern enrolled boys, ages five to twenty-one, in a program of lessons, field trips, and farming experiments based on research at the college.

By 1905, the club had more than five hundred members. Kern waxes rhapsodic about the club's power to interest the boys "more deeply in the beauty of country life and the worth, dignity, and scientific advancement in agriculture," but the boys' real source of motivation for joining was probably financial: corn and sugar beet experiments yielded real harvests, and Kern reports that the club members earned about $15 each in 1905—almost $400 in 2013 dollars.

If the boys' motivation was cash, then the agriculture college's was public relations. Kern explains: "The growing of high-bred corn by the boys is a movement to get both them and their fathers interested in improved types of grain." In the years that followed, agricultural colleges organized contests for boys to see who could grow the most corn. One awards ceremony dinner in Nebraska featured an all-corn menu, which included

Figure 1. A girls' canning club in Nebraska, 1919. From *Junior Soldiers of the Soil* (Des Moines: Meredith Publishing Company, March–April 1919), 27.

corn soup, corn relish, corn bread, corn pudding, and even a popcorn float.

Kern's, Otwell's, and Graham's efforts eventually collided and merged. The club-and-contest model worked so well that agriculture researchers began to use it to disseminate other farming improvements. In 1908, the USDA hired a former South Carolina superintendent of schools named Oscar B. Martin to expand the network.[7] As the clubs spread, their focus widened. In Iowa, rural teachers taught their students how to test the butterfat content of milk. In some places, teachers started girls' clubs for sewing, baking, and, most importantly, canning. It was around this time that researchers were developing safe methods for putting up food, and girls' clubs were key to promulgating these new best practices. Martin worked tirelessly on the canning clubs. Wessel and Wessel describe how, in the summer of 1910, he dragged a tin can canner wherever he went for "canning bees," where girls competed to can the most tomatoes.[8]

In the South, agriculture professors seized on the club model as a way to encourage farmers to diversify their fields. Cotton had been the only cash crop in the entire region for so long that the soil was beginning to suffer. Through the clubs, the researchers launched a campaign to sell southern farmers on corn, which could be used as food for both humans and animals. To demonstrate corn's value as a livestock feed, a land-grant college in Mississippi launched the first pig clubs.[9]

The job of heading all the clubs became too big for Martin alone, so in 1910 O. H. Benson, an Iowa superintendent, joined him. Benson is best known for developing a partnership between federal and local governments to fund the clubs, but he was also responsible for publicizing 4-H's clover logo, which had been designed by a schoolteacher named Jessie Field Shambaugh, with whom Martin had worked in Iowa.[10] The four leaves of the clover stood for the four H's: head, heart, hands, and health. By this time, some members were already referring to the clubs as 4-H, though the name wasn't officially adopted until 1924.

In 1914, the federal government passed the Smith-Lever Act, which created the Cooperative Extension System, a new partnership between the federal government and universities.[11] The club model had proven that rural farmers could benefit from scientific research in college agriculture departments, and Congress had taken notice. With the passage of the Smith-Lever Act, universities began to receive federal and state funding for their work with farmers, of which the clubs were a major component. With this switch, Extension agents from the university—many of whom traveled for miles to rural areas—replaced superintendents as club leaders.[12]

Club membership rose steadily following the passage of the Smith-Lever Act, and when the United States went to war in 1917, demand for food for the troops skyrocketed, and even more children joined clubs to contribute to the effort. According to Wessel and Wessel, 169,000 children participated in clubs in 1916. Just two years later, more than 500,000 did.[13] Livestock clubs, formerly relegated to the South, took off nationwide.

A few weeks after my conversation with my neighbor Kristen, I attend a meeting of the Montclair 4-H club at a community center up in the Oakland hills. It's controlled chaos: teenagers limp through Robert's Rules of Order, and as the evening wears on the littlest members—kindergartners and first graders, known in 4-H as "Cloverbuds"—start to lose it, whining to their moms and provoking one another. The group skews young, as does 4-H in general: it serves mostly kids ages five through nineteen, though there are also collegiate clubs. In 2011 membership nationwide peaked at close to a million in fourth grade and gradually declined as kids grew older; just 130,000 4-H'ers were high school seniors.

The items on the agenda tonight are mostly club business: fundraisers, carpools, snacks. Hardly anyone even mentions animals, and I begin to feel antsy, too. When the teenagers finally, gratefully, adjourn the meeting, I introduce myself to a few parents, most of whom live in Montclair and nearby neighborhoods. I ask whom to talk to about the goat project. Everyone says the same thing: you must meet the Hawkey sisters.

I learn that the sisters in question, Chloe and Serena Hawkey, are two of the teenagers who were leading tonight's meeting. Chloe, a seventeen-year-old high school senior, is the club's president, as well as the teen leader of both the goat and rabbit projects. Her fourteen-year-old sister, Serena, is vice president of membership. Both have been in the club since kindergarten—and, in fact, their parents helped found the club. These are girls who take their animals seriously, other parents say in reverent tones. They win prizes at the fair.

I e-mail Chloe and Serena's mother, Andy, thinking I'll be able to catch up with the girls in the week before the next 4-H meeting, but it turns out that most of their evenings have been spoken for. Serena has cross-country meets, Chloe has college applications to work on, and both girls have homework and hours of ballet classes. We make plans to meet two weeks later. I invite Chloe and Serena out to Fenton's, a celebrated Oakland ice cream parlor.

When the evening for ice cream comes, I drive up through affluent neighborhoods in the Oakland hills, past mansions and vast yards. Finally, I pull onto a narrow street that is so densely lined with trees it has a woodsy feeling about it. The Hawkeys live in a modest split-level set into the side of a hill.

Andy opens the door and greets me warmly. She is slight and sporty with straight brown hair and a friendly smile. In the homey living room, a 4-H skit project meeting is just wrapping up. A handful of elementary school kids race around, and Chloe and Serena and a few other teenagers pore over a binder. Parents trickle in, staying for a minute to chat, then cajoling their kids out the door.

After a few minutes, Chloe and Serena extract themselves from the group to say hello to me. Serena is unmistakably Andy's daughter—same long, straight hair, same athletic build. Chloe is tall and willowy, with an oval-shaped face framed by shoulder-length light brown hair. The girls have excellent manners for teenagers: strong handshakes, good eye contact. Despite the fact that I have never met them before, they don't seem at all shy. They say goodbye to Andy and we head out to my car.

Minutes later, we sit down at a table at Fenton's. As we stare down enormous dishes of Heath bar and peanut butter cup, the girls talk not about animals, but about school.

"Homework," groans Chloe. "I have so much homework."

"Me, too," says Serena. "I still have a test to study for when we get home."

"I gather that school is really important in your family," I say.

"Yes," says Chloe. "But our parents aren't the pressure kind of parents. It's more like we put pressure on ourselves." Their dad, she says, often finds her starting her homework at 10:30, after an evening full of activities. "He goes, 'Isn't it a little late to be starting your homework? What would happen if you just didn't do it? Take the night off and get some sleep?' And I'm like, 'Dad, I don't know what would happen if I didn't do my homework, but I don't want to find out.'"

Both Chloe and Serena have GPAs above 4.0. They attend a magnet program at Oakland Tech, a public high school that draws kids from all over the city. Only a few of Chloe's and Serena's friends from school are members of 4-H. The rest, who fill their out-of-school time with sports, drama, and community service, occasionally express mild curiosity about the club. Someone recently told Serena that in middle school she was known as "goat girl," which she found funny.

"So is it nice to go to the fair every year, where people understand what you're talking about when you mention 4-H?" I ask.

Chloe and Serena look at each other.

"Kind of," says Serena. "But we're not like the other clubs." The county that the Oakland club belongs to, Alameda, is sprawling, she says. Out toward its northern and eastern borders are real agricultural communities. "Culture out there is very different from Oakland culture," adds Serena. "The girls out there listen to country music and wear pink cowboy boots."

"It's not that we don't like them, or that they're not nice," says Chloe. "But when it comes to 4-H, they definitely have an advantage. The main difference is that they have access to all things rural." Sometimes, says Chloe, she envies their feed stores, land, and family and friends who have been farming and ranching for generations. "We can do the things that take dedication," she says. "But not land."

"But obviously they don't have too much of an advantage," I say. "You guys win awards."

"We are extremely competitive," admits Chloe. "But just in showmanship."

Chloe explains that there are two kinds of competitions at the fair. "One is about how good the animal is," she says. "With dairy goats it's called conformation. With animals that are sold at auction—like pigs or sheep or steer—it's called market. For dairy goats, it's mostly the formation of the udder. For meat animals, it's their balance of muscle and fat. Other stuff, too. Like bone structure. To do well, you have to be extremely careful about how you raise and exercise the animal. You also have to have an animal with really good genes. Which means getting it from a breeder or working on breeding the right kind of animal on your own farm. That can take years."

"Showmanship is more about the kid," Serena chimes in. "How much control you have over your animal, and how much you know about it. The judge asks you questions. You have to study. We're really good at studying."

"Studying is definitely where we excel," agrees Chloe. "Some people might think it's weird, but I actually enjoy the process of studying for Rabbit Bowl."

"Rabbit Bowl?" I ask.

Chloe explains that it's a quiz-bowl-style competition that takes place at the fair, where contestants answer questions about rabbit trivia. The Montclair senior team—composed of the group's high schoolers—has never lost.

"My best friend and I study a lot for it," she says. "We love doing nerdy things like that."

In fact, says Chloe, she even wrote her college application essay about Rabbit Bowl. She hopes it will catch the attention of the admissions committees at the liberal arts schools she's applying to. I can imagine administrators at Vassar, Amherst, and Middlebury enthusing over this high-achieving kid with an unusual hobby.

We finish our ice cream. I have more questions, but it's getting late, and Chloe and Serena have to start their homework, so I drive them home.

Over the course of several visits to the Hawkeys' house, I gradually learn more about Montclair 4-H. Andy and Steve Hawkey, a financial planner and college professor of finance, respectively, founded the club with a few other parents when Chloe was in kindergarten and Serena was in nursery school. Andy had heard of 4-H—in fact, two of her nieces out on the coast were members—but she was fuzzy on the details. One of the other moms told her that 4-H allowed kids to decide what they wanted to study and also to run the meetings themselves. That idea appealed to Andy. She did a little more research, and she liked what she read, especially the emphasis on "learning by doing." She and Steve liked the animal component, too. Steve had chickens growing up and wanted to try goats. Participating in 4-H would be the perfect way for the whole family to learn how to take care of them.

The Hawkeys and a few other Montclair parents recruited the club's first members through word of mouth and a few fliers. Steve Hawkey and another club dad led a poultry project, and other mothers and fathers offered cooking and sewing groups. Meetings were small and casual; the club had fewer than twenty members. After several years, the parents decided that Montclair 4-H was ready to go to the fair. "It was obvious we were green," says Andy. "We were city kids, not country kids. But people were very welcoming." After the club members showed their chickens, the judge called all the Montclair parents over for a talk and told them that their kids had been using the wrong kind of chicken. Show chickens are typically bantams—a smaller variety that elementary school children can easily manage. Montclair 4-H'ers were using full-sized chickens. Chloe, who was by then six, had a hard time keeping her huge chicken, Pepe, under control.

"The judge gave us a talking-to," says Andy. "He told us that we had to get the kids animals that they actually could handle. We felt so bad."

Still, the Hawkeys were taken with the scene at the fair. Serena remembers idolizing the 4-H kids that she met. "I really wanted to be like those other kids at the fair," she says. "I wanted to be a cowgirl."

In fact, all four Hawkeys were in awe of the 4-H champions—kids who had learned to show sheep and steer from their parents and grandparents.

They were fascinated by the farm jargon they overheard in the barn. What would it be like, they all wondered, to live on a working ranch? Andy and Steve imagined doing barn chores instead of office tasks, spending weekends at livestock auctions, and riding horses into the sunset. "We all got cowboy boots and hats," remembers Andy. "We were a complete wannabe family." I could relate.

The club grew over the next few years. The Hawkeys decided that they wanted to add a goat project, so they partnered with another club mom and bought two Nigerian Dwarfs—a dairy breed popular among 4-H'ers because of its small size and relatively cooperative nature—and kept them in their small yard. The Hawkeys fell in love with these two curious wethers (neutered males), but the creatures proved to be deft escape artists, and it became clear that they'd need a dedicated space. Andy placed an ad in a neighborhood newspaper. Right away a woman answered. She owned a property up in the hills that had a yard overrun by weeds and brush. "She told us that if we could get the goats to eat all this brush, we could use the place for free," says Andy. The club transformed a few shacks in the messy yard into homes for the goats. They've lived there ever since.

Another Montclair family started a pig project in a similar way. At first, the pigs lived at one family's house, but as the project membership swelled, they outgrew the close quarters and were relocated to a shared space near Oakland's city-owned horse stables. Today the goat and pig projects have fifteen to twenty members each. Montclair 4-H has about a hundred members—the most it's ever had. Andy attributes the growth in part to 4-H's unique hands-on learning model, which she thinks appeals to parents. But she believes that the urban farming trend has helped, too. "Some of the parents are interested because they know it's a way that they can be involved in this farming thing," she says, "even if they don't live out in the country."

Andy is proud of the club's success, but she still sees room for improvement. She notes that the club is "very white." (That's true of 4-H in general:

nationwide, about three-quarters of members identify as white.)[14] Most of the Montclair members are from upper-middle-class families. For a while, an Oakland city councilman showed interest in recruiting kids from other parts of the city. A few families joined, but it was often hard for parents to schlep their kids all the way up to the hills several times a week. Andy would like to figure out a way to make it easier.

Andy still devotes hours every week to 4-H, but her daughters are now busy teenagers, and the club is only one of their interests. A few other families have stepped into leadership roles, a development for which Andy is grateful.

I'm curious about the Montclair dairy goat herd ("the goaties," as the Hawkeys call them), so I arrange for a visit. Again, it's hard to schedule a time when both girls are free. Finally we give up on Chloe, whose weekend is jammed with ballet classes. It'll just be Serena, Andy, and me.

On a crisp Sunday morning in November, I ascend once again into the hills. The property is near a network of regional parks. A few spandex-clad cyclists whiz by me on my way up. After a tough climb, they'll be rewarded with sweeping views of the Bay. Most of the homes in this well-to-do neighborhood are meticulously maintained, with tasteful landscaped gardens. When I arrive at the address that Andy has given me, I see that the goats' home is different. The rambling old wooden house is in desperate need of a paint job, and scotch broom is colonizing the sloping yard.

Andy opens the gate, and we head down a path to a small structure, where Serena is sweeping the floor. Once she's done, I follow her to the yard out in back of the shed, where seven or eight goats contentedly munch on fresh grass. A few playfully butt heads. Their coloring varies; some are whitish with bands of honey-colored fur around their middles; others are darker all over. Nigerian Dwarfs are technically dairy goats, explains Serena, though the club doesn't milk them regularly. Nevertheless, at the fair, they'll be judged on the quality of their udders.

Serena recites the litany of their names so quickly that I only catch about half, but I notice an Italian theme: Perrugia, Fiera, Bianca, Valentina. "My family lived in Italy for a year, so we were giving them all Italian names for a while," she explains.

The group keeps food, medicine, and cleaning supplies in a closet attached to the shed. The goat duty schedule is posted on a calendar. The families in the project take turns letting the goats out in the morning, closing them back up at night, feeding them, and cleaning the pen. Goat duty typically takes between twenty minutes and an hour, depending on how messy the pens are.

We watch the goats nibble grass for a while, and then Serena has to return to her homework, so we part ways.

Over the next month, I begin to attend Montclair 4-H goat meetings. The kids' attention spans are short, but Chloe and Serena press on with their goat lesson plans. At one meeting, they divide the members into small groups and hand out crossword puzzles with clues having to do with goat trivia. I visit a pair of third-grade boys who seem to be immersed in their puzzle— but when I get close enough to hear their conversation it becomes clear that they are actually just retelling the plot of *The Empire Strikes Back*. Chloe comes over to give them a few hints. Another day I watch Serena teach a group of squirmy grade schoolers an easy trick for memorizing parts of a goat's back— withers, chine, loin, rump: "Winning champions love root beer," she says enthusiastically. "Say it with me now." She seems surprised and pleased when some of the kids actually do.

Chloe and Serena are 4-H outliers in that they are livestock-raising members of an urban club. Today, most 4-H'ers come to the organization through its classroom, after-school, and camp programs. In fact, only a quarter of this generation's 4-H'ers belong to clubs. And even in clubs, not everyone raises animals. Members are busy building model rockets,

Figure 2. Serena Hawkey, 14, works with Bianca, one of the Nigerian Dwarf goats from the Montclair 4-H herd. Photo by Rafael Roy.

Figure 3. Chloe Hawkey, 17, with Kajsa, another goat. Photo by
Rafael Roy.

preparing skits for drama night, learning about water conservation, or pursuing any number of other projects.

I was impressed that the Montclair group had found a way to raise animals in Oakland. But I was also curious about the other 4-H'ers that Chloe and Serena told me about—the kids who raise superior goats because they have "access to all things rural," as Chloe put it. I still wanted to meet one of those kids—the farm girl or boy who learned to show livestock from her parents and grandparents. I wanted to find the 4-H'ers who had made the Hawkey family all want cowboy boots and hats.

I would have to keep looking.

CHAPTER TWO

Learning by Doing

I ask around and learn that Pleasant Hill, a suburb just northeast of Oakland, also has a 4-H club with livestock projects. So on a cold January evening, I hop in my car and head east, through the tunnel under the Oakland hills. In about half an hour I arrive at Foothill Middle School's combination cafeteria-auditorium for the Pleasant Hill 4-H Club's first meeting after Christmas break. Right away, I notice two things:

1. All of the thirty-odd kids seem to be wearing pajamas.
2. A woman who appears to be dressed as some kind of crustacean is standing near the stage.

I'm a few minutes late, and as I take my seat in the back, the kids are already standing to recite the 4-H pledge:

I pledge my head to clearer thinking
My heart to greater loyalty
My hands to larger service
And my health to better living
for my club, my community, my country, and my world.

The kids settle back into their seats, shuffling their fuzzy slippers and re-arranging their bathrobes. The lady crab minces out to the front of the room and introduces herself.

"Hi everyone," she says. "I'm the Collaboration Crab." She gives an enthusiastic talk about how the group will have to work together to prepare for the club's upcoming fundraiser, a Dungeness crab dinner. Next to the crab lady, a few older girls sit behind a folding table. One of them, in a pink bathrobe and lots of eye shadow, alternately smiles tolerantly at the Collaboration Crab and looks down at her lap to text.

After a few minutes, the texter calls the meeting to order. "Thank you guys for dressing up for pajama day," she says. "Everyone looks awesome."

The rest of the meeting proceeds according to Robert's Rules of Order, just like the Oakland meetings I've been to. The treasurer reports the club's income since the last meeting: $73. Members from several projects—scrapbooking, cooking, rabbits, goats, and photography—give quick reports. There's a lot of talk about fundraisers; one mom announces that she plans to crochet a basket that looks like a chicken for a live auction. The last order of business is a raffle for candy. All the kids have drawn tickets on their way in, and two lucky ones come forward to claim their prizes.

At the end of the meeting, I wait to talk to the adult leader of the club, KC Chatham. Although she is not wearing pajamas like the kids, she looks comfortable in a sweatshirt and jeans, her hair in a long braid down her back. She addresses kids and parents alike in an informal manner that seems to put everyone at ease. She smiles often. During a break in the steady stream of people who are approaching KC with questions, I introduce myself and ask if it might be possible to talk to some of the kids. She says I am welcome to talk to her sixteen-year-old daughter, Allison Jefferson, the one in the pink bathrobe who was leading the meeting. Allison is the club president.

I follow KC over to where Allison is talking to another teenage girl. Allison, I see up close, has long, dark hair and big, expressive eyes set off by thick black eyeliner. In a quiet voice, she tells me that she raises rabbits and two pygmy goats. She invites me to visit the goat club, which meets at Old Borges Ranch, a former working cattle farm not far from the middle school. Later that week, KC sends me directions in an e-mail, which is signed:

KC Chatham~
Pleasant Hill 4H Community Leader
(_ _ /)
(= '.' =)
(")_(")

Borges Ranch turns out to be a 1,475-acre property in the foothills of Mount Diablo, one of the highest peaks in California's East Bay. Purchased by Francisco Borges in 1899, the heyday of the cattle industry in Northern California, the ranch was home to hundreds of head of cattle for decades. The Borges family sold the land to the city of Walnut Creek in 1975, but they leased the ranch back and continued to prosper until a drought in the late 1980s forced them to scale down. They sold all the remaining cattle in 1991. Today, Borges Ranch is a registered historical site, with a preserved redwood cabin, barns, a blacksmith shop, antique farming equipment, and sprawling pastures bounded by valley oaks. The city lets Pleasant Hill 4-H keep its animals there for free, as long as the club keeps the pens clean.

I head out to the ranch on an unseasonably warm January day. It's just nineteen miles from Berkeley, but I feel like I'm out in the country. From a small parking lot, hiking trails lead off into the rolling hills. A few turkey vultures soar overhead. The grass is a bright emerald green from the winter rains.

I walk up a path and poke my head into one of the old, high-ceilinged barn buildings. On the walls hang old-fashioned tools—rusting saws and axes with warped wooden handles. Clustered around a hand-hewn table is the goat club—four teenage girls, including Allison. KC is leading the lesson.

"If you guys see the judge checking the goat's teeth, what do you do?" she asks.

"He's not allowed to do that anymore," says Allison, "since it spreads Johne's disease."

Once KC is done with the lesson, the other girls introduce themselves: Sydney, Nicole, and Erika, all sixteen. In addition to this crew, KC tells me, there are two younger girls, sisters named Alyssa, thirteen, and Kayla, eleven, who couldn't make it today. All the girls around the table are wearing makeup and eyeliner. They all have long, shiny hair, which they wear loose and toss often. Their nails are flawlessly manicured. KC claims that the day's task is to vaccinate the goats, but the club looks ready for an afternoon at the mall.

We head outside. I follow the girls along a path to a large gated enclosure. This, Allison explains, is where the goats live; the girls take turns doing the daily feeding and cleaning, just like in the Oakland club. Allison points to her goats: Choncho, a brown-and-white four-year-old wether, and Rayne, a pretty eight-year-old doe. The other two are Nicole's all-black doe Midnight and Erika's wether Chewy. Next month, says Allison, two baby goats will join the herd—Sydney and Kayla have decided to raise their own goats this year.

The girls catch their goats, put them on leads, and walk them back up the path. I can't help laughing at the pygmy goats shuffling along on their leashes. A full-grown pygmy goat is only as tall as a medium-sized dog but weighs about eighty-five pounds. They seem to carry it all in their bellies, which, according to the judging standards, is a good thing: the National Pygmy Goat Association describes the ideal belly as "broad, deep, increasing in width and depth toward flank, thus giving an impression of perpetual pregnancy." They remind me of hobbits: very round and slightly annoyed at being made to stop eating. I sort of expect them to have squeaky little cries to match their physique, but when one of them lets out a "bah," he sounds like a little old man complaining.

Back outside the barn, Allison expertly pulls Choncho onto her lap while Sydney uses a syringe to inject medicine into the indignant goat's mouth. Allison explains that they're vaccinating the herd against overeating

Figure 4. Allison Jefferson, 17, and her pygmy goat, Rayne, at Borges Ranch in Walnut Creek, California. Photo by Rafael Roy.

disease, in which a dietary change causes normal bacteria in the animal's gut to proliferate. Allison directs Sydney, who seems tentative with the syringe. "You have to be fast about it," Allison says. "Otherwise he's just going to spit it out." Sydney gamely pries open the goat's jaws. The two girls finally manage to get the medicine into Choncho's mouth, and he scrambles off Allison's lap, looking thoroughly dejected. The girls repeat the process three more times for the other goats, with KC offering suggestions and encouragement from the sidelines. Amazingly, Allison's and Sydney's manicures emerge from this ordeal unscathed.

After the vaccinations are done, it's time for the girls to do some cleaning and feeding. I offer, hopefully, to watch a goat or two while they're working.

"Technically, you're not supposed to," says Allison. "In 4-H, the kid is supposed to be the only one handling the animal, unless it's an emergency." I try not to look disappointed as the crew heads back to the barn.

I hang back and talk to KC. As we chat she puts the vaccine supplies away and gathers up the papers from today's meeting, her braid swinging behind her. She tells me that she used to work as a vet tech, but she's been out of a job for a few years. She's lived in Pleasant Hill since she was three years old. Her father worked at his father's lumberyard in a nearby town, and her mother, who used to run a home day care, is retired. KC lives in the house she grew up in, with her mother, Allison, and her other daughter, Lexi, Allison's half sister. Lexi's father lives nearby, but Allison's lives in Los Angeles.

In addition to being the goat project leader, KC heads the club's rabbit project and costume design project. She's also the community club leader, which means that she coordinates monthly all-club meetings and fundraisers. Over the years, she's worked with "literally hundreds" of 4-H'ers.

"I like teaching them," she says. "And I like seeing their faces when they win."

KC didn't mean to be the leader for so many projects, but the other parents quickly learned that she is the kind of person who sees something that needs to be done and does it. The former rabbit leader got busy and asked KC to take over a few meetings. The same thing happened with the goat leader. And pretty soon, the other leaders had dropped out, and KC was running the show. (Although the Extension—in California, it's the University of California, Davis—employs 4-H youth development specialists, the actual meetings and projects are typically led by trained volunteers, often parents.)

The girls finish their barn chores and the meeting wraps up. On our way back down to the parking lot, Allison mentions her 4-H rabbits and chickens, which she keeps at home. I tell her I'd like to meet them, so we make plans for me to come out to Allison and KC's house in Pleasant Hill, a few miles away from Borges Ranch.

The following week I set out one evening after work to visit Allison and KC. Since I know that Allison raises animals at home, I imagine her neighborhood with farmhouses and barns and rambling pastures. But when I pull off the highway, I'm on a wide, busy street lined with strip malls and fast-food restaurants. I make a left that looks like it will lead me straight into the parking lot of a Grocery Outlet, but instead it takes me behind the store into a cute neighborhood of tightly packed houses. I pull up to one with a flag out front, and before I can even park, KC and Allison emerge. Apologizing, they tell me that Allison's grandmother thinks the house is too messy for a visitor and ask if we can go somewhere else. I agree, but I am disappointed—I had wanted to meet Allison's animals. We decide to go to a Starbucks, where we order hot drinks, settle down at a table, and start talking about Allison's 4-H career so far.

"It was my older daughter—Allison's sister, Lexi—who did 4-H first," says KC. "We went to a breeder to get her a rabbit when she was ten." The breeder recommended 4-H. KC had heard of the group, but she "didn't even know 4-H existed here." But she found the local club and enrolled Lexi in the rabbit project. After about two years, a couple of the girls in the club got pygmy goats. "Another girl wanted to get a baby, and it's always better to get two," KC says. "So she asked Lexi if she wanted to get one. We decided to do it."

Even before Allison was old enough to join 4-H, says KC, Lexi taught her how to show rabbits and goats. "I always wanted to be like my sister," Allison tells me. "So when I was five I joined this 4-H program for little kids called Pocket Pet. I raised a hairless rat." The next year Allison got her first rabbits and chickens to show at the fair. In 4-H, she explains, you can show small animals—rabbits, chickens, guinea pigs, and a few others—when you're as young as five, but you have to be nine to show large animals. Allison didn't have her own goat until she was twelve, but she showed her sister's goats long before that. She was a natural from the first time she set foot in the show ring. Rarely has she come home from a fair without a prize; she has boxes of ribbons and trophies from goat, sheep, rabbit, and chicken shows.

At first, when Allison was just starting out, she found competitions stressful, especially the part where the judge asks questions about a species' anatomy, health, and breeding. "I used to study for weeks before the fair," she says. "My mom would quiz me all the time, and I would learn from the older kids."

One year, she decided she wanted to show chickens. None of the other kids in her club knew how, so Allison had to teach herself. She and KC got their hands on a poultry handbook, and Allison studied every night. When the fair came, she took home first prize in showmanship. The next year, she started a chicken project and taught the younger kids everything she knew. She found that teaching the material helped her nail it down. The only problem was that for some of Allison's students—such as her thirteen-year-old neighbor and fellow 4-H'er, Alyssa—her lessons worked too well. "I taught her everything she knows about rabbits, and she beat me at fair last year," Allison says with mock indignation. "She got second place for showmanship, and I got fifth."

For the past few years Allison has had her heart set on winning a competition called "masters"—small-animal master showmanship. I've heard about this competition from Chloe, who won small-animal masters last year at the Alameda fair. (Allison does not compete against Chloe and Serena, because Allison's club belongs to Contra Costa County 4-H, which has a separate fair.) First-place showmanship winners from each small-animal species compete against one another. Each contestant has to show an animal of every species—in the case of small-animal masters, that's a pygmy goat, a rabbit, a guinea pig, a chicken, a pigeon, and a dog. Large-animal masters includes a meat or dairy goat, a sheep, a steer, a pig, and a horse. Allison has never won masters. Last year she missed a few easy questions during the rabbit showmanship contest, because, she says, the judge was really cute.

Hearing about Allison's success in 4-H and all her studying, I suspect that, like Chloe and Serena, she excels academically. But when I ask "How's school?" her whole manner changes. Speaking about 4-H club, Allison has been lively and forthcoming, but now her expression is blank.

"It's okay," she responds. "My school is really big." Last year, when Allison was a sophomore, there were sixty students in her math class. That was a problem, since Allison needed extra help. She has trouble focusing in class, and she sometimes has a hard time remembering directions that her teacher has given. Oral presentations are the worst. A few weeks ago, she was supposed to give a talk for English class on Martin Luther King Jr. for black history month. Allison was so nervous that she got sick and had to go home. She never ended up giving the presentation.

KC says that Allison's problems at school started suddenly when she was in fifth grade, right after she suffered a bad concussion as a result of a playground accident. Her vision quickly declined around the same time, too, and Allison has worn glasses ever since. Doctors haven't been able to say for sure whether the concussion caused Allison's problems. She has been tested for learning disabilities, but she scored just above the range that would qualify her for accommodations.

Allison's school has all the usual cliques, each of which, Allison says, stakes out its own claim in the cafeteria. At the first four tables are the football players and the cheerleaders. The dancers sit in a corner, and the drama types and potheads hang out everywhere. It is not the kind of place where you are cool if your activity is 4-H. Allison's best friend, Sydney, one of the other girls from the goat group, goes to a different high school.

"Only one more year of high school after this one," says Allison. "That's how I like to think about it."

I can see that school isn't one of Allison's favorite things to talk about, so when we finish our coffees, I decide to call it a night.

As I begin to attend more Pleasant Hill 4-H meetings, I watch Allison closely for signs of the problems that she and KC told me about at Starbucks—difficulty focusing, trouble remembering instructions, nervousness about public speaking. But that version of Allison is hard to reconcile with the one that I am getting to know—the confident, knowledgeable, and outgoing 4-H Allison. In the cozy old barn at Borges Ranch I watch her explain goat care to the other girls. Later, she and Sydney go down to the goat enclosure to muck the stalls, seemingly unconcerned about getting goat poop on their carefully constructed outfits. I appreciate their stylish approach to goat management.

Then one day, about a month after my first visit to Borges Ranch, KC tells me that Allison has enrolled in a new independent study program at school called Horizons. Instead of attending classes, Allison meets for a few hours each week with her Horizons advisor and then works through her assignments at home. KC has high hopes for the program, since it allows Allison to teach herself and work at her own pace. It's going well so far—Allison beams when she tells me she received a 93 on a history test. She has never gotten anything higher than a 78 in history before.

"It's weird, because I'm really just teaching myself in this, but I'm doing better," she says. "I can go in to meet with my teacher and spend time with him. Plus, I have friends who have already taken the same subjects in this program, so they can talk me through it."

"That sounds like 4-H," I say. "Kids teaching other kids."

Allison thinks about this. "I learn more from actually being shown how to do something. If someone tells me and takes me through it, then I work a lot better than if I sit in a class and listen to someone talk about it."

A few weeks later, I discover that Allison is not the only one who struggles at school while excelling in 4-H. A leader of Lamorinda 4-H, a nearby club in the suburb of Martinez, California, tells me that there is another member I must meet. His name is Anthony Cannon. He is only in eighth grade but has already taken home top honors in swine market and showmanship competitions. Although pigs are Anthony's passion, says his club leader, he attends every 4-H competition that he can—public speaking, baking, and even fashion revues.

I get in touch with Anthony, who sends me the following information about himself:

> I am thirteen years old. . . . I am raising two market show pigs and a market lamb. I will be exhibiting the lamb and pigs at the Contra Costa County Fair. One pig will come home and be shown at the California Youth Fair in late June.
>
> I have shown animals ever since I was nine years old. I have earned several showmanship awards. And three grand championships on my projects.
>
> I raise my animals at a ranch about a mile from my house. In Martinez. The barn and property are owned by former 4-H leaders who are retired. About a year ago I wrote a letter to them to see if it was possible for me to raise my animals on their property.
>
> My future goals include being a firefighter and involved in animal science.

Anthony's mother invites me to come out to the property where Anthony raises his animals. I find the place easily—it's an old farm lot right in the middle of a residential neighborhood. I walk up a long, sloping pasture, past three miniature horses, to a barn and adjoining outdoor animal pens at the top of the hill. Rusting tractors and other old, broken farm equipment lie around.

A middle-aged woman with curly brown hair and a shy smile introduces herself as Susan, Anthony's mother. "Anthony's in the barn," she says. She calls into the barn for him to come out. Anthony emerges briefly, waves, then ducks back in. Susan sighs. "Nope, Anthony, come back out and talk for a minute," she prods. And then, to me, she says, "It's hard to get him to focus on anything else besides the animals when he's here."

While we wait for Anthony to finish whatever he's doing in the barn, Susan tells me that the people who own this property are old friends of her family; Susan grew up doing 4-H with their son. The couple now rents out the small house south of the barn, but as for the property itself, "they weren't using it for anything, so they were delighted to have him here," she says.

Anthony visits his animals every day after school during the week. Usually he spends about an hour cleaning their pens and feeding and exercising them. When fair time approaches, he increases his time here to two or three hours a day. When she has time, Susan drives Anthony to the animals, but more often, he comes by himself on his bike or scooter. Susan manages a local Chuck E. Cheese, and her husband, Anthony's dad, is a freelance TV news videographer. They work a lot, so Anthony's grandparents, who live nearby and also have been involved in 4-H for most of their lives, take Anthony to 4-H events.

Finally, Anthony emerges from the barn. He's a stocky kid with tousled hair and different-colored rubber bands on his braces.

"Hi, nice to meet you," he mumbles. Two small pigs—about the size of corgis—clatter out after him, running around the pasture, looking happy to be free. He's leading a sheep on a leash.

"Explain to Kiera what you're doing," says Susan.

"I'm exercising them."

"Tell her why exercise is important," she presses.

"Because you need them to have muscle at the fair. That's what the judge is looking for."

We watch the animals run around. I meet Satchel, the sheep, whom Anthony has had for about a month. The pigs, an energetic pair, are Panda, a female who was born in December, and Domo (named for a cartoon character featured on Anthony's hat), who was born around November. Anthony just brought Domo home from a breeder yesterday. Panda and Satchel each cost a few hundred dollars, but Domo, Anthony tells me, was free, the result of a scholarship from the California Pork Producers Association. Anthony will enter Domo and Satchel in the county fair in a few months, but Panda won't be big enough to qualify by then, so he will enter her in a different competition a few weeks after that.

I ask how he finds the money to pay for his animals. "When I was nine, I earned enough money for my first pig by collecting cans. I made a lot of money on her, so I used my earnings to buy two animals the next year. I've been doing that ever since."

"My other big thing is public speaking," he continues. He's won prizes for 4-H presentation day, most recently at a national competition. I raise my eyebrows. This kid has to be prodded to answer my most basic questions, and I'm supposed to believe he's a public speaking champion? I ask him what he likes to speak about. The speech he wrote for the national competition, he says, was about the *Wizard of Oz*—how Dorothy's farming background "kept her on the straight and narrow."

"I talked about how if she drank alcohol and took drugs, she wouldn't have stayed on the yellow brick road. One of the judges was teasing me after. He said I got it all wrong, that in the real ending, the scarecrow got burned up from smoking a cigarette, the tin man got all rusted out because he spilled beer on himself, and you don't even want to know what happened to Dorothy."

"Ha," I say. "I would have loved to have seen that."

I start to ask Anthony whether he's working on any speeches now, but he has already turned back to the animals, supervising their playtime. After Satchel tires out, he hands her leash to Susan. Then he grabs a long plastic stick. He uses this to round up the pigs, gently tapping their sides to show them where to go. He explains that you have to use a light touch with the stick. Hitting the pigs with it is mean and unethical, and, furthermore, it could bruise the most valuable cuts of meat. Swine judges can tell right away if pigs in the show ring have marks from owners who are overzealous with the stick, and they deduct points for this.

"See Panda, the little one? She thinks Satchel, the sheep, is her mom," says Anthony. We watch Panda follow Satchel around for a while, the piglet sniffing hopefully at the sheep's hooves. Domo keeps to himself. "He's more wild, because he came from a big pig farm and isn't used to being walked."

Then it's time to weigh the animals.

"Anthony is very meticulous about keeping records of their weight," says Susan. "He wants to make sure they're on track to make weight for the fair, and also that they aren't sick." Susan tells me that Anthony does not love academics, and he struggles especially hard with math at his small Catholic school. However, when it comes to using his animals' weights to calculate the amount of food they need, he "does just fine." "The teachers recently figured out that if they relate the math problem to his animals, he can do it," Susan says. I'm reminded of Allison.

Satchel and Panda step onto the livestock scale without a fight, but Domo, the new pig, squeals and kicks before Anthony has even gotten him in the barn door. Even as the pig tantrum escalates, Anthony stays calm. "Oh, quit your whining," he says as he corrals Domo. "You're okay."

Figure 5. Anthony Cannon, 13, bonds with his pig Panda at the barn space he borrows in Martinez, California. Photo by Rafael Roy.

When the drama is over, he lets the pigs into their pen and sprays with a hose to make them a mud wallow. The pigs scamper over and hunker down in the damp.

Once the animals are safely inside the pen, Anthony sits right down next to them in the mud and begins to scratch the pigs' ears. Susan catches my eye and gestures to Anthony as if to say, *Look at my stoic and taciturn teenager sharing a tender moment with his swine!*

Had I found the authentic 4-H'ers I was looking for in Allison and Anthony? I wondered. I had seen their bona fides: Allison vaccinated goats; Anthony carefully recorded his lamb's and pigs' weights and was even a third-generation 4-H'er. Both of them lived a little farther out of the city than Chloe and Serena. But their hometowns, full of big-box stores, chain restaurants, and side streets leading to subdivisions, were not exactly farmland—they were suburbs.

If I had undertaken this project eighty-five years ago, I wouldn't have had to venture far from Oakland to find farm kids. From the archivist at a local historical society, I learned that in 1930, if I had stood facing west on top of Mount Diablo, the highest peak in the area, I would have seen several cities: Berkeley and Oakland due west, and a little farther south, a smaller one called Hayward.[1] Farms would have stretched out around the cities. The biggest wouldn't have been more than several hundred acres, with a modest house and barn set into a shady corner. If I had climbed down from the mountain and walked the fields, I would have seen orchards—cherry trees in the north, almond and apricot trees in the south, and pear and plum trees scattered throughout. In the city of Hayward, I would have seen Hunt's Cannery, where trucks brought tomatoes for canning every fall. The strong odor of tomatoes, the historian told me, overwhelmed the town from August through October.

Although most of the farms grew one or two crops to sell, each had a large vegetable garden where the farmer's family grew its own food:

peas, fava beans, rhubarb, strawberries, tomatoes, cauliflower, cucumbers, sugar beets, squash, and currants. Most households kept chickens for eggs, along with a few other animals: a dairy cow or two, maybe a few sheep.

Few of these farmers were rich, but almost all lived comfortably, with enough food for the family and a little extra cash to spend at the county fair, the highlight of every summer. A good number of the farm kids raised animals to show at the fair, many through 4-H. All the teenagers attended Hayward Union High School, where many joined the thriving chapter of Future Farmers of America, logging hours every afternoon at the school's farm.

The population of the Hayward area grew from about seven thousand in 1940 to seventy-two thousand in 1960. During the intervening decades, the landscape changed dramatically. Developers bought up the orchards and fields and converted them to subdivisions, wooing potential buyers with the promise of safe, quiet neighborhoods just a short drive away from Oakland and San Francisco. By 1957, with more people settling in south Hayward, the school district gave up Hayward Union High School's Future Farmers of America plot in order to build a new high school. In 1981 Hunt's Cannery, which had operated for a century, moved its operations to California's Central Valley, as did many of the other nearby canneries and farm-related businesses. Today, the vast majority of farming in California takes place in the fertile Central Valley and its surrounding areas. Only a few working farms are left in the East Bay, and most of their owners have another primary source of income.

I was beginning to realize that if I wanted to find a farm-country 4-H'er, I would have to go to farm country.

CHAPTER THREE

"I Do Sheep the Way Other Kids Do Soccer"

Having made up my mind to look for 4-H'ers in California's farm country, I decide to visit the Greenfield 4-H club in the fertile Salinas Valley, which begins inland from Monterey and stretches ninety miles down to just north of San Luis Obispo. Flanked by two mountain ranges—the Saint Lucias to the west and the Gabilans to the east, the region was made famous as the setting of many John Steinbeck novels. Today it produces more than $4 billion of fruits and vegetables every year, including 80 percent of America's lettuce.[1] Greenfield, a town of sixteen thousand in the southern part of the valley, is the self-proclaimed broccoli capital of the world.[2]

I make the trip down to Greenfield one Sunday afternoon, heading south past Fremont and San Jose. Just south of Santa Cruz, the landscape changes. Fields line the roads. First strawberries appear, and then, farther south, past the city of Salinas, the pungent smell of broccoli fills my car.

The sun is setting. The crisp greens and yellows of the fields fade, and the oranges of the earth come out. In the distance, the Gabilan Mountains are a deep violet. Produce trucks whoosh past me. Most, laden with empty crates, are likely coming back from dropping off their vegetables at distribution centers. In the last few minutes of daylight, I pull off the highway onto a stretch of California's historic El Camino Real, which serves as the main drag of Greenfield. Lining the street are markets, cafés, and a few storefronts advertising services through which people can send money home to Mexico. All the signs are in English and Spanish. I know from reading about Greenfield that 90 percent of the town's residents identify as Hispanic, and many of them work in agriculture. I duck into a *tortilleria* to see if I can buy some tortillas to take home, but the guy says he doesn't sell fewer than a hundred at a time, too many even for someone who likes tortillas as much as I do.

I make my way over to meet Randy Sosa, the seventeen-year-old president of Greenfield 4-H. A coordinator at the Monterey County 4-H office has told me that Randy and his siblings are some of the most active members of their club, never missing a meeting. Randy's neighborhood reminds me a little of Allison's, with small suburban homes built close together. The Sosas' house is a one-story with a well-kept yard. I ring the doorbell. A wiry teenage kid with close-cropped dark hair opens the door. He smiles broadly, showing his mouthful of braces.

"Hi, I'm Randy," he says. "Come on in." I follow him through the front hall into a cheerful, well-lit kitchen. Randy's dad comes in from outside. He is not much taller than Randy, but stockier, and with the same wide smile (but no braces). When he shakes my hand, I see fading tattoos covering the inside of his arm. "I'm Eleazar Sosa," he says in heavily accented English. "Welcome to our home."

Randy, Eleazar, and I sit down at the kitchen table, and Randy's younger siblings—Roosevelt, sixteen, and Jocelyn, fourteen—come in and sit on stools at a counter. Instead of a tablecloth, a piece of glass covers a poster of the periodic table of the elements, made, Eleazar explains, by his wife, Alba, who hopes that the kids will learn something while they're eating. Alba earned a degree in chemical biology in Mexico and was a teacher there. She became a nursing assistant when she came to the United States.

"She knows everything about homework," says Randy. "She enhanced her English skills just to help us."

Eleazar works in the vineyards. He came to Greenfield in 1978, after studying veterinary science in Mexico. "I was looking for jobs here caring for animals," he says. "But my English was not very good, so I wanted to go back to school. Then I came into the grapes. I don't know if I'll go back to school now. In the grapes I have worked a long time, so I have a lot of responsibilities."

"Pretty much everyone is in agriculture here," says Randy. "Almost everyone's family in my school works in the fields."

"Do you want to go into agriculture?" I ask.

"No, I want to be an engineer," says Randy. "But my brother likes agriculture."

"Yeah," says Roosevelt, turning to us from his stool. "But I don't want to pick grapes or broccoli. I want to be in the corporations."

Eleazar nods approvingly. Even though he likes his work, he says that he hopes that 4-H will prepare his children for careers beyond the Salinas Valley's fields.

Randy tells me that he has been in 4-H for eight years. He and his siblings mostly do woodworking and rocketry projects, but, like Anthony, they also compete in public speaking contests. Last year, Randy's presentation, which was about how to fix a flat bike tire, made it all the way to the statewide competition.

"It helps to practice with family," says Jocelyn, joining the conversation. "We do our presentations here in the kitchen. The rest of our family asks questions so we can get prepared for the judges."

Randy has never done a livestock project. He says he has always wanted to—having a sheep or a pig, he thinks, would be really cool. But his family doesn't have enough room at home. "And we do other stuff besides 4-H, too," says Randy. "We do cross-country, track, basketball, and we teach karate in our garage."

"In your garage?" I ask incredulously.

Figure 6. From left to right: Roosevelt, Randy, Eleazar, and Jocelyn Sosa at their home in Greenfield, California. Photo by Kiera Butler.

"It started with my older brother. He became a second-degree black belt and was certified to teach. He started teaching our cousins. But then people started hearing. Now we have thirty-two students."

I ask to see the studio. Randy opens a door off the kitchen that leads to the garage, which is indeed outfitted as a makeshift karate studio, with benches, mats, and mirrors on the walls.

On our way back to the kitchen, I notice a wall of family pictures, medals, and plaques. Randy points out his older brother, Eleazar Jr., in his ROTC uniform. "This wall is inspiration for us," says Randy. For college, so far Randy's first choice is the Air Force or Naval Academy. Eleazar tells me that Randy is ranked fifth in his class of 211 students.

"I am proud," Eleazar says. "But also they are not always here at home. When they were little, we did things as a family, dinner together every

"I Do Sheep the Way Other Kids Do Soccer"

night. But now one goes to cross-country practice, another to 4-H, another has karate. We are never all home at the same time. But it's good. I tell them, participate. If you are busy, then you don't have time for anything bad."

We tour the rest of the Sosas' small house. The kids show me their woodworking projects and the model rockets they built and painted for the fair. Just as we're finishing our conversation, there's a knock at the back door. Eleazar answers and welcomes in a young woman with two little kids—a boy and a girl—shyly clinging to her. She and Eleazar exchange a few words in Spanish, and Randy tells me that this is the 4-H leader, Natalia. He introduces us, and Natalia sits down at the periodic table.

Natalia tells me that she grew up in Greenfield on a ranch. After graduating with a teaching credential in agricultural science, she found her current job in the agriculture commissioner's office in Salinas. This is her first year as a volunteer 4-H club leader. She invites me to the club meeting the next day, telling me that she has lined up a special guest: a pumpkin farmer. "He's going to talk about what it's like to grow crops, and how different seeds turn into different pumpkins, and that kind of thing," she says.

This strikes me as strange. Given that most of the kids in this community are from farming families, I ask, don't they know firsthand what it's like to grow crops? Natalia explains that while most of the kids have a parent who works on a farm, few of them live on farms. "They don't see what the work is all about," she says. "The parents' dream is always, 'My son or daughter is going to be better than me. They're going to be an engineer, a lawyer, a doctor.' Rarely do parents want their children to come back into agriculture." Natalia tries to show the club members all aspects of agriculture, from the basics of crop production to the wide range of careers in the industry.

The 4-H meeting is at the Lions Club in downtown Greenfield. Randy, Roosevelt, and Jocelyn—all club officers—are at the table up front, and kids are filtering in, finding their seats. All but a few of the thirty-five or so kids present have dark hair and eyes, and they speak with their parents in a mixture of Spanish and English. In the other 4-H clubs I've visited, I've seen only a few nonwhite faces. In fact, Allison, whose father is black, and her friend Sydney, whose mother is Mexican, are the only 4-H'ers of color I've met so far.

Today, 75 percent of 4-H'ers identify as white, 15 percent as African American, 15 percent as Hispanic, and 2 percent as Asian (members can identify as more than one race). Because clubs tend to mirror the demographics of their larger communities, they are often homogeneous. Greenfield's club is composed mostly of Hispanic members because the town's population is predominantly Mexican American.

Buried in 4-H's history, though, is a much more troubling racial dynamic. In 1948, a 4-H foundation was established to help with organization and fundraising. One of the foundation's first victories was raising the money to lease and eventually buy a former junior college in Chevy Chase, Maryland. The campus would be transformed into a national 4-H center, which would house trainings, meetings, and other important 4-H events, including an annual conference and camps.

During this era, if you worked hard enough at your 4-H projects, you could earn a trip to this fancy new headquarters—that is, unless you were black. Like schools, 4-H programs in the South were segregated. Black 4-H clubs had limited funds, and, in most places, they weren't allowed to compete with white clubs. There were separate awards programs; when black 4-H'ers won prizes, their names weren't included in official 4-H publicity materials like white kids' names were. And while white club members could work toward attending national meetings in Chevy Chase, black kids were not allowed.

Carmen Harris, a professor of history at the University of South Carolina, has studied the discriminatory policies of the Extension toward black farmers in the South. Harris believes that the main goal of the Extension was to turn farmers into businessmen. But that wasn't true for the black laborers, who had been working the land since the days of slavery. The Extension "perceived black farmer entrepreneurs as a threat," Harris told me. "They had a lot of subversive potential." This attitude extended into 4-H clubs. In a 2008 paper, Harris describes how in 1913 the early 4-H leader Oscar B. Martin specifically recommended against allowing black girls to form canning clubs.[3] "Martin insisted that African American girls should not be encouraged to get into the 'business of canning,' presumably because it promoted self-sufficiency," writes Harris.

The deep divide between white clubs and black clubs lasted for decades. Harris relays a story told by a black journalist named James L. Hicks, who in 1946 asked Gertrude Warren, a prominent Extension leader who is sometimes called the "mother of 4-H," why black children weren't allowed to attend the camps. Harris quotes Hicks describing his interview with Warren: "She said she was New England-born and 'did not believe in discrimination,' but that 'our southern [directors] . . . are very strong about that sort of thing. If we invited our colored 4-H workers to attend the convention with the white, we wouldn't get a delegation of white southerners.'"

In 1948, in response to an outcry over the discriminatory policies of the annual camp, the Extension grudgingly opened a regional camp for black 4-H'ers that continued for the next twelve years at various locations in the South. D. W. Watkins, an Extension official in South Carolina, wrote to the regional camp leaders that they were not to spend "public tax money for the purpose of bringing in Negro celebrities or any other celebrities, except as they may make a contribution to the basic purposes of holding the camp." Even in the late 1950s, as the rest of the United States was waking up to the idea of civil rights, southern 4-H leaders remained staunchly committed to segregated clubs. The Extension defended those clubs, arguing that doing so was a matter of honoring states' rights.

In 1955, a year after the *Brown v. Board of Education* verdict, the Extension considered integrating the annual camp in Washington, DC, but federal 4-H director E. W. Aiton argued in a letter to the US secretary of agriculture that "emotions and commitments are such that the State workers would send an all Colored delegation rather than face their own [white] people after sending a mixed delegation." Although individual leaders voiced their objections to the policy of the regional camps, it wasn't until the Civil Rights Act of 1964 that the Extension was forced to integrate all 4-H activities—and even then, instead of integrating, some southern states chose to close their 4-H clubs and camps. They didn't open again until several years later.

Today, it's hard to find any evidence of this dark time. The history section of the National 4-H Council's website does not mention club segregation, nor do any other 4-H historical materials that I found. I can understand why the organization would want to forget this shameful chapter of its history, but Harris assured me that some older African American 4-H alumni remember it quite clearly.

The Greenfield meeting goes pretty much like all other 4-H general club meetings I've attended: little kids squirm in their seats as teen leaders soldier on through minutes and bylaws. I am beginning to develop a distinct distaste for Robert and his mind-numbing Rules of Order. Randy does an admirable job trying to keep the kids engaged, and he handles a cranky parent calmly and confidently, all while grinning disarmingly. The highlight of the meeting is the pumpkin farmer. He has brought with him five or six pumpkins ranging from apple sized to five hundred pounds. He explains that each pumpkin was grown from a different seed, and he has brought seeds with him to hand out. The kids love the idea that the seeds for tiny and humongous pumpkins look so similar.

After the pumpkin farmer leaves, Natalia, who, in addition to being the club leader, also runs the sheep project, announces that she is trying to find a central space where the kids can keep their lambs. Like the Sosas, she explains to me after, most of these families don't have enough space to keep animals; while the parents work on sprawling farms every day, they return home at night to modest houses with small yards.

At the end of the meeting, I say goodbye to Randy, Roosevelt, and Jocelyn. Randy thanks me for coming, and we agree that the pumpkin farmer stole the show. He invites me to come back and visit again.

On the long drive home, it's too dark to see the broccoli fields on either side of the road, but I can smell them. I have definitely found farm country—and yet, despite being from farm families, Randy and the other Greenfield 4-H'ers don't have any more access to farmland than Chloe, Serena, Allison, or Anthony.

I want to take Randy up on his invitation to come to Greenfield again, but the next time I e-mail him, he answers from Rhode Island, where he is attending the Naval Academy Preparatory School. I can imagine how proud the rest of the Sosas must be.

Just when I have begun to believe that the authentic 4-H'er might be a figment of my imagination, a Montclair parent tells me that there is a livestock-raising club that I should visit: Palomares 4-H, which meets in Castro Valley, California, a suburb about half an hour southeast of where I live. I e-mail with the club leaders and am invited to attend an evening meeting. I find the house where the meeting is taking place at the bottom of a steep, narrow driveway off a dark, sparsely populated road.

A mustachioed man wearing a sweater and jeans answers the door, introduces himself as Steve Semonsen, and leads me into the kitchen, where a group of parents chats around a woodstove. The house is spacious but cozy, with rough wooden beams, high ceilings, and homespun decorations. It reminds me of the houses featured in the *Country Living* magazines my mom subscribed to when I was little.

Steve directs me to a room in the back, where about a dozen kids sit in folding chairs arranged in a circle. A rabbit hutch is set up in the corner.

Plaques line the walls: Champion wether dam. First place, breeding ewe. Prize ribbons hang from a set of coat hooks. At the front of the room is a dry-erase board sitting on an easel. It reads:

Palomares 4-H Club
GOALS:
- Champion lambs
- Belt buckle
- Lots of money
- Fun

A teenage girl holds a clipboard in her lap. I suspect that this is Kelly, Steve's daughter. As I take a seat, she announces that a member named Kyle will talk about what buckets to use for feeding and watering. A small, skinny kid shuffles to the center of the circle and motions to a pile of buckets. He picks up the containers one by one and, in a series of Hemingwayesque sentences, explains how to use each one: This one is for feed. You put the food in it. That one is for water. And so on. Next up is a middle-school-aged girl who answers a series of questions about exercising lambs.

"You have to exercise your lamb so it builds muscle," she says.

"And?" prompts Kelly.

"And, um, lambs have a lot of energy!"

"So how do you exercise your lamb? Do you just let it run around?"

"No, you have to halter train it."

"Right. And which should you do first—feed your lamb, or exercise it?"

Hands shoot up. Kelly calls on one of the little girls.

"Exercise it?"

"Right," says Kelly. "Because if you feed it first, it could get cramps when it exercises, just like when you eat a big meal, then try to run around."

The rest of the meeting goes on like this, with Kelly patiently extracting information from members of the group and filling in the gaps when the presenters miss something.

When Kelly adjourns the meeting, I'm full of questions, but I have to wait my turn; she's busy doling out advice to the younger members and some parents who have come to pick up their kids. By the time I corner her it's getting late.

Kelly, who is sixteen, has an all-American look about her—long blonde hair, a petite frame, and serious blue eyes. Answering my questions, she is poised and articulate, but not chatty. She tells me that she has been in 4-H for thirteen years, and that her main focus in the club has always been lambs. I ask her how many sheep she has. She scrunches up her nose and reflects for a minute.

"About thirty?"

"Wow!" I exclaim. "All for your 4-H projects?" Kelly laughs and explains that her family raises sheep, both for other kids to show and for meat. She says that the plaques on the wall are mostly hers, though some belong to her fourteen-year-old brother, Kyle, the boy who gave the presentation on buckets. I ask Kelly about the plaques, and she gives me a tour of her past victories. She tells me that she is more competitive in showmanship than in the market competitions. She's qualified twice for large-animal master showmanship.

"I got stepped on by a steer in the ring the first time," she says. "Then the second time went better. It's tough to learn how to show animals you're not familiar with. You have to ask your friends. You have to be like a sponge and absorb it."

She hopes to win this year, though as a high school junior she has less time than ever to practice with her lambs because of increasing amounts of homework and SAT prep. Right now she works with her sheep for about an hour and a half every day, a little more on weekends and right before shows. "I do sheep the way other kids do soccer," she says.

In addition to county fairs, Kelly says, she goes to other lamb competitions called jackpot shows. Whereas at county fairs, participants make money only from auctioning their animals after the competition, at jackpot shows, kids can win prize money for their animals. These shows are much more competitive than county fairs, says Kelly. "You pull in with your trailer, then have fifteen minutes to get all your gear out and disinfect your pen. Then you get your lamb out and give it a bath and a haircut. Then you check in. It's very tense." Kelly tells me that some jackpot show participants hire professional adults to train their lambs. "Then the kid just shows up to the show with the lamb." But Kelly doesn't see the point in hiring someone else to do all the dirty work. She always does everything herself.

Unlike jackpot shows, fairs don't allow any adults to help train kids' lambs. In fact, the only adults involved are the judges, the auctioneer, and the auction attendees who buy the animals. I ask Kelly how the auction works.

"You want to make sure you have a few possible buyers in advance," she says. "That's actually the hardest part of the whole project." Every year, Kelly and Kyle write letters to potential buyers, all of which she has saved. She runs to retrieve them and comes back with what looks like a bunch of Christmas cards. She lays them out in chronological order. In the first one, a few sentences of text are accompanied by several pictures of ten-year-old Kelly, looking very small compared to some of the sheep. The letters get longer and more sophisticated each year. I read through one that Kelly and Kyle sent three years ago:

Hello from Kelly and Kyle Semonsen,

We are members of the Palomares 4-H Sheep Project. I am 14 years old and this will be my 5th year showing my bred-by-exhibitor lambs at the Alameda County Fair. Kyle is 11 years old and this will be his second year showing his bred-by-exhibitor lambs.

Last year we purchased a new ram for our ranch. We named him Monty. He did a very good job breeding our ewes and we know you will be pleased with the quality of our lambs this year. It was really hard to choose which ones we wanted to take to the fair! Anyway, we have been working very hard

"I Do Sheep the Way Other Kids Do Soccer"

raising our lambs from birth. They require shots, a special diet, and lots of exercise. Together we have been preparing our lambs for show and last year Kyle took 1st place in showmanship. We even had 4 of our lambs make the Championship drive.

We think you will like our lambs. Both Kyle and I will be showing 3 to 5 lambs each in the Market Class (I told you we couldn't choose) on July 5, 2009, at 10:00 am and you [can] see us in the Showmanship Class on July 7th at 2:00 pm.

I especially hope you will come to the auction where we will be selling 4 of our lambs on July 12, 2009. We can get you all the information you need to register to become a buyer. We bet our lambs will taste good too!

Hope to see you there!

Kelly and Kyle Semonsen

Kelly explains that she has built up a stable base of repeat customers. "I usually ask the guy from a local pool company, and the guy who owns the Safeway," she says. "They all come to the auction on the last day of the fair. People take their lambs up in the order of how well they did in the ring. There is sometimes a lot of crying. The boys are hardest to say goodbye to, but they can't stay, because they can't do anything."

The Semonsens need only one breeding ram for all the ewes on the farm, so ram lambs, Kelly explains, are castrated when they're very young—after which they're known as wethers. Champion ewes can come home to breed more champions, but young males must eventually be sold.

From the two male lambs that Kelly auctioned off at the fair last year, she made a total of $2,000. After she paid back her parents for feed and other expenses, she still had $1,600 for her college fund.

Kelly considers the fair—and all sheep shows—career training. All her life, she has wanted to be a large-animal vet. She has it all planned out: she enrolled in her high school's nursing program, which she hopes will prepare her for the college biology classes she wants to take. Next year, she will apply to the College of Agriculture at the University of California, Davis, where she will major in animal science. Then she'll apply to stay on as a graduate student in the school's highly regarded veterinary medicine program.

We head back into the kitchen, where I meet Kelly's mother, Teri, who is efficiently tidying up as the last of the 4-H parents depart. She has shoulder-length brown hair, a round face, and a demeanor that is friendly but focused, like Kelly's. We sit down at the kitchen table, and she tells me that her husband's parents established the farm in 1968 and raised sheep for decades. When they passed away, Steve and Teri took over the farm. The lambs are not their main source of income, however. Steve is a building estimator, and Teri works in a school cafeteria.

Teri invites me to watch the family's two border collies herd the sheep—eighteen lambs and twelve adults—into the barn for the night. Teri trades my clogs for a pair of rain boots and Kelly joins us as we head out into the pasture. The field is much bigger than I had thought, and in the corner I can just make out a mass of gray—the sheep.

"Okay, Stella, get 'em!" yells Teri.

The border collies are off and running. In an instant, the gray mass is in motion. Sheep of all sizes scatter.

"Easy, Stella, easy!" cries Kelly.

The dogs circle the herd in a wide arc around the field, and the sheep charge toward the door of the barn. For a split second it looks like they are coming our way, moving so fast that I'm scared. But the dogs are on it. They steer the herd into the barn, where there is much jostling and pushing as dozens of sheep try to fit through the door at one time. The whole process has taken about forty-five seconds.

Once the commotion is over, I follow Kelly into the barn. She scoops up one of the tiniest lambs, a flailing, crying gray guy, no bigger than a cat. He's about two weeks old. "He still has his long tail now, but we'll dock it soon," says Kelly. I ask why. Fair regulations, she explains. Sheep that are shown must have exactly enough tail to be lifted with a pencil.

Figure 7. Kelly Semonsen, 16, offers a snack to two lambs on her family's farm in Castro Valley, California. Photo by Rafael Roy.

Teri shows me the rest of the barn, including the area where two pregnant sheep live. "We keep a baby monitor in here so we'll know when the action happens," she explains. "Then Steve comes out to make sure everything goes okay."

Kelly leads me over to her two show lambs. They wear red jackets, looking very important. While I am petting one, another little lamb sneaks up behind me and begins chewing on my pants. It is easily the cutest thing that has ever happened to my pants.

I could stay in the barn with the show lambs and the babies and the rest of the bah-ing group all night, but it's getting late. We head back into the house.

As we're walking back from the barn with the dogs, it dawns on me: Kelly is the real 4-H'er that I've been looking for. I suddenly wonder if she owns pink cowboy boots—that hallmark of real 4-H'ers that Chloe and

Serena told me about—but Kelly is just so serious that I can't quite bring myself to ask about something so trivial.

Teri and Steve won't let me leave until they've given me two frozen lamb chops. I look at the meat, then at Kelly.

"Did you, um, know this lamb?" I ask awkwardly. Kelly smiles.

"I probably did, but it wasn't a 4-H project," she says.

I make plans with Kelly to come out again soon and drive back to Oakland.

A few days later, some friends and I thaw the lamb chops and cook them for dinner. We panfry them simply with salt and pepper. Despite the time they have spent in the freezer, the chops taste incredibly fresh—rich and grassy.

My quest could have ended there. In Kelly, I had found what I was looking for: she had grown up on an actual sheep farm, and she lived and breathed 4-H and lamb shows. But in the weeks that followed, I wondered about the urban and suburban 4-H'ers and their animals—Chloe and Serena up in the Oakland hills, Allison, whom I had watched vaccinate her pygmy goats, and Anthony and his pigs. Even if they didn't match my original image of 4-H'ers, they were certainly having real agricultural experiences. What were they learning from 4-H? And what would they do with all their specialized knowledge about raising animals?

And then KC, Allison's mom, e-mailed to tell me that two members of the Pleasant Hill goat group—Sydney and Kayla—were planning on picking up new baby goats from a breeder. Did I want to meet them? Oh yes, yes I did. I wrote back immediately. I assured her that I would be at Borges Ranch—the property where I had watched Allison and Sydney give the goats their shots—the moment the babies arrived.

CHAPTER FOUR

Bringing Up Baby

On the Sunday that the baby goats are scheduled to arrive, I get to Borges Ranch about half an hour before everyone else. It's an unseasonably warm February day—bright sunshine, birds rioting in the blue oaks, cows grazing on the hillside. A minivan finally pulls up, and all the members of the goat group—Allison, Sydney, and Erika, along with the younger girls, Alyssa and Kayla—tumble out of the car. Allison and Sydney are holding two toy-poodle-sized baby goats, both black with white faces. They're ten weeks old, says Allison. Axel, the bigger of the two babies, belongs to Sydney. Butterball, the smaller one, is Kayla's. KC grabs a few leashes from the barn.

"Remember, they're not used to being on leashes yet," Allison reminds her mom.

Fastening the leashes onto the baby goats involves a lot of shrieking and giggling. When Sydney and Kayla finally manage to do it, we all watch the babies skitter around. It's hard to tell whether they are terrified or just exuberant. Sydney and Kayla have their work cut out for them. The fair is still more than three months away, but the baby goats are completely untrained.

And neither girl has had a goat before. Sydney wanted to get one last year, but she didn't have enough money. This year her parents consented and each chipped in. She named her goat Axel after a dog in the movie *Million*

Dollar Baby. Sydney is a little worried about the logistics of caring for Axel; her parents don't have time to drive her to Borges Ranch. She hopes that her boyfriend, who has a truck, will help.

Sydney and Allison have been inseparable since they met in health class in seventh grade. Both girls love animals. In the apartment where Sydney lives with her mom, who runs a cleaning business, she has a rat, a guinea pig, and a rabbit. Her Mexican grandparents live on a ranch in the country-side in Zacatecas. Sydney loves visiting them and spending time with their cows and sheep.

Eventually Sydney wants to become a veterinarian—but first, right after she graduates from high school, she plans to enlist in the US Army Reserves. "I have always wanted to do something for my country," she tells me. "And I think the structure will be good for me." As I got to know Sydney more, I could see her excelling in the kind of high-stress situations she might encounter in the army. She hardly ever spoke up during goat meetings, but it was clear that she was listening. More than once, when something chaotic was happening—a goat off its leash, or Allison struggling to hold her goat and grab feed at the same time—Sydney was there, quietly doing exactly what the situation called for.

The girls, the goats, a few parents, and I head down to the goat enclosure and let Axel and Butterball off their leashes. A troop of boy scouts wanders by, and the boys are delighted by the goats. They have a lot of questions: How long will the goats live? Will they ever be able to see their mother again? Allison answers patiently.

Even though Butterball is Kayla's goat, Kayla's older sister, Alyssa, seems to be doing most of the leash handling. At thirteen, Alyssa is less than two years older than Kayla, but while Kayla still looks like a little kid—long, messy blonde ponytail, loose sweatshirt, and jeans—Alyssa is starting to dress more like the older girls. Alyssa clearly idolizes Allison—she's always trying to be close to her, bumping into her deliberately, pulling on her scarf.

Allison is very tolerant, as well as quite comfortable bossing Alyssa and Kayla around.

The girls start to clean out the stalls for the new goats. The babies will live by themselves for a few weeks, in case they've brought a disease with them from the breeder that could spread to the rest of the herd. As the girls sweep, there is lots of prom talk. KC is making dresses for Allison and Sydney out of some hunting camo fabric. Both of them love the idea of making a statement with their prom dresses instead of just buying something conventional from the mall.

Once the sweeping is done, Allison and Sydney head up to the barn to fetch hay and alfalfa for the new goats. Kayla gets sidetracked playing with Butterball.

"Kayla!" yells Allison. "Get your butt up here and help us carry hay down for your new goat!"

Inside, Kayla gapes at a poster about goat vaccinations.

"You're going to have to do that, so you better get used to it," says Allison.

"Gross, even the one in the udder?" asks Kayla, pointing to the picture of a needle going into a doe's teat. Allison reminds Kayla that her goat is male, so he doesn't have udders. "Giving them shots is nothing compared to deworming, though," says Allison. "The babies shake their heads around and try to spit the medicine out." Kayla's eyes widen. "Don't worry," Allison reassures her. "You'll get it."

When the girls have filled a wheelbarrow with supplies for the new babies, we take it to the babies' new home. The girls fill the water troughs and put down fresh hay, alfalfa, and feed. By the end of the afternoon, both babies are shivering out of nervousness and exhaustion. When the girls pick them up they immediately relax, and Butterball looks like he is ready to go to sleep. Kayla is reluctant to leave her new goat.

"What if he gets lonely?" she asks.

"He'll be fine," says Allison.

Figure 8. Allison Jefferson (left) and Sydney Williams with their goats at Borges Ranch in Walnut Creek, California. Photo by Rafael Roy.

"Ask your dad to take you up here tomorrow, but if he can't, you call me so I know to get someone else to come feed them, okay?" says KC. Kayla nods.

Most Northern California 4-H'ers have decided on animals to take to the fair by the end of winter. Many choose to get new animals, either by buying them from a commercial breeder (as Sydney and Kayla did) or by breeding the animals they already have. There are advantages to both approaches: If you breed your own animal, you're eligible to enter it in a special bred-by-exhibitor competition at the fair. But breeders offer more variety; a 4-H'er can select from the specific characteristics offered.

At the end of April, Chloe and Serena Hawkey's mom, Andy—leader of the Montclair 4-H goat project—mentions to me that they think that three of the group's goats are pregnant. A few months back, the club paid a breeder for "stud service"—meaning they brought their does to visit a buck for a few days. While Andy and the goat project members don't know for sure which of their goats were romanced by the stud, they have been monitoring all the females to see which ones are putting on weight. They are almost certain that Valentina and Kajsa are pregnant; they're not as sure about Bianca, who might just be getting chubby. I tell Andy to keep me posted on any signs of labor. I'd like to see a goat birth, but Andy warns me that things can happen pretty quickly.

A few weeks later I receive an e-mail from Andy. The ligaments around Valentina's tailbone are very soft—a sure sign that labor will happen within the next day. I try my best to make it to the birth at the goats' enclosure in the Oakland hills, but I'm coming from more than an hour away that night, having attended an evening lecture at Stanford. By the time I make it back, to my dismay, I've missed the whole thing. Luckily, the morning after, Serena offers to give me a minute-by-minute account of the action.

On the evening of the goat birth, Chloe had ballet practice, homework, and college applications to fill out, so Andy dropped off Serena by herself. She found Valentina lying down in a corner of the yard. Luckily, Serena had been present for a few goat births before, so she knew what to expect. But she had never been alone with a laboring goat. She sat beside Valentina, who licked her hand. This is something that goats do when they're in labor—they become very affectionate.

Serena felt around Valentina's tailbone for the ligaments. They were gone, so she looked for signs of contractions. But nothing seemed to be happening. Valentina seemed agitated and kept licking Serena's hand.

And then, all at once, Valentina contracted. Her whole back end scooped under, and her tail lengthened, elongated by the engaged muscles. She stood up and began to rub herself on the fence. The contraction ended, but about three minutes later another one came. Pretty soon they were coming every two minutes, then every minute. Valentina wandered around the yard, unable to get comfortable in any one place. Serena followed calmly. She spoke to Valentina in a low voice and petted her when she looked uncomfortable.

About two hours in, Valentina lay down near the fence and Serena saw her lip curl under—a sign that she was about to push. Serena ran up to the barn to retrieve the kidding supplies: an aspirator, wipes and iodine, Vaseline, gloves, old towels, and a pencil and paper to record the timing of the progress. But when she got back, she became worried: the ground by the fence was covered in rocks, some of which were sharp. The baby could come any minute, so she had to act fast.

She picked up the straining Valentina and carried her up to the stall. Valentina lay down, nickered, and began to push. Her mucus plug started to come out, which meant that her water was about to break. Serena laid out a towel, and Valentina turned herself around and pressed her two back hooves against Serena's legs. A bubble emerged from Valentina, and after a few seconds, it burst. Serena got a new towel. Valentina pushed again, but nothing came out.

Chloe arrived just as Valentina started pushing again. Finally, after another hard push, a little bit of tan-colored fur emerged. It was a snout, which was followed by a little mouth, with a tongue sticking out. This was not ideal; goats are supposed to be born with their front hooves first, in a Superman pose; otherwise the hooves can become stuck inside. Valentina pushed some more, and she began to bleed. Something inside her had torn.

For what seemed like forever only the nose and mouth were showing. They were full of mucus, which Serena suctioned out using the aspirator. She couldn't tell whether the umbilical cord was still attached; if it wasn't, then the baby could have drowned in mucus.

Valentina gave a big push, and one hoof emerged—which meant the other one was curled under by the baby's side. This explained why Valentina was having such a hard time pushing. Serena and Chloe debated reaching in to get the hoof, but they had never intervened in a birth before, and they didn't want to risk hurting the baby. They decided to let Valentina push for a few more minutes. If nothing happened soon, they would call the vet.

Valentina pushed some more, and the baby's head and torso appeared. This was a relief; Serena and Chloe could pull the baby out the rest of the way if they needed to. Normally, at this point, the goat would turn around and the baby would emerge fully, breaking the umbilical cord. But Valentina seemed exhausted. She lay still, panting.

And then she started pushing again. Another baby goat slid out all at once, over the top of the first one. His umbilical cord broke, but the first baby was still halfway inside Valentina, so Serena pulled him the rest of the way out. He lay there and blinked. Serena and Chloe cheered him on, telling him to break his umbilical cord. After about twenty minutes, he did.

Valentina licked both babies all over, and after just a few minutes they tried to stand up. At first, their legs gave out from under them, but eventually they were able to wobble around. Their hooves were soft and spongy; they wouldn't harden for another few hours. The babies were very skinny—all bones and ribs. They needed to nurse right away. Serena picked one up and tried to put him on Valentina's teat, but he couldn't quite find it. He tried to suck on her armpit instead.

The first feeding was extremely important, because Valentina was producing colostrum, a substance that contains essential nutrients for the babies' first days of life. Her body would make it only for a few hours. Finally, with some help from Serena and Chloe, both babies found the teats. Their little bellies became fat. At around 11 P.M., the babies and Valentina lay down, exhausted. Serena was exhausted, too, but also ecstatic.

The morning after the birth, I drive up to meet the two new babies. Serena and Andy lead me down to the goat pen. One of the babies is sitting with his hooves tucked under him, like a figurine in a Christmas nativity set. The other is standing next to him, looking like he owns the place, despite being less than a day old. They are the same color as the hay surrounding them. Valentina is nearby, too. Her back end looks raw, but she doesn't seem to be in pain. The standing baby ambles over to her, and she gives him a few licks.

"Do you want to hold one?" asks Serena.

Um, *of course I do*. She gathers the sitting one and hands him to me. He is about the size and weight of a petite cat. He nestles placidly into my arms.

Serena and Chloe have decided to name this baby Huckleberry and the other one Finn. The babies were born just in time for them to qualify to be shown at the fair. It was hard to imagine Sydney's and Kayla's new ten-week-old goats being trained in time for fair, but the idea of these even tinier goats in the show ring just seems ridiculous, like sending an infant off to college. Serena assures me that they will grow fast.

I let Huckleberry go, and he wanders over to Valentina to nurse.

"So, how was it being here on your own yesterday?" I ask Serena. Delivering a baby goat seems like a lot of responsibility for a fourteen-year-old.

"It was cool," says Serena. "There were a few minutes when I was nervous. But mostly I knew what to do."

As I learn about the process of bringing the newest members of the Montclair goat herd into the world, I begin to notice all the small extra expenses—beyond just feed and bedding—that go with raising animals: Stud service. Vet bills. Aspirators for delivery. And it isn't just the babies that require additional stuff. At one Castro Valley lamb project meeting, Kelly tells her fellow members that even if they've raised the best lambs in the world, all kinds of things could go wrong at the fair, and they must prepare. Water purifiers are a good idea, since the water at the fair is different from what the lambs are used to, and they can be finicky. Blankets are essential for keeping lambs clean before they enter the show ring. Kelly's mom, Teri, leans over to me and whispers, "We just bought three new lamb blankets. A hundred bucks. Can you believe it?"

One of the things that initially appealed to me about 4-H livestock projects was their emphasis on entrepreneurship. I loved the idea that a kid could carefully raise an animal, auction it off, and end up with a nice little nest egg.

I should have known that it didn't always work out like that—if for no other reason than the enormous expenses that I accrued raising my turkeys. I bought lumber for their shelter, feed and watering systems, hundreds of pounds of organic meat-bird food (which I probably could have just eaten myself, skipping the turkeys altogether), and dozens of cans of tuna. My friends and I hemorrhaged cash on those two birds. I am not sure we could have spent more money if we had tried, unless we had actually sent them to Montessori preschool. If we had sold them, we might have been able to extract $100 for each one, on account of their privileged upbringing. We still would have been at least $1,000 in the hole.

The first expense of a 4-H livestock project is the cost of the animal itself. This year, Kelly bought one moderately priced show lamb from a breeder and selected a few others from her family's flock. Allison helped Sydney and Kayla choose a reputable and affordable breeder for the baby pygmy goats that I met at Borges Ranch. For the 4-H'ers I've met so far, fancy breeding is not a top priority, since they are most interested in showmanship. They know that with enough hard work, they can train almost any goat to obey them in the ring.

For 4-H'ers who are serious about market classes, however, the right breeding often makes the difference between winning a ribbon and going home empty-handed. And for those who don't live on farms with breeding operations, animals with champion lineage come with a hefty price tag. Anthony's pig-project leader, Sally Pereira-Cox, warns the members of her group that if they spend exorbitant sums on their pigs, their profit margins will be thin. Most members of her group spend around $300 on a pig, but she has heard of kids spending $1,000 or more per animal. Anthony was lucky to receive scholarship money to buy his pig Domo, whose lineage is excellent. His younger pig, Panda, is not as fancy, but she was relatively inexpensive.

Sometimes a big initial investment pays off; a supreme champion hog, lamb, or steer can fetch a much higher price at auction than an animal that doesn't place. But the strategy of blowing all your cash on a fancy show animal doesn't always work. Sally points out that even an animal with an excellent pedigree and top-of-the-line accessories could end up getting sick and not making the fair's weight cutoff. Or what if the judge at your show just happens to prefer a different style? Then your pig doesn't place and doesn't do as well at auction—and you're out hundreds or even thousands of dollars.

An additional cost for all livestock is feed. Show feed is usually much more expensive than commercial feed, since it contains more supplements and a higher percentage of protein. Tiffany Burrow, the Alameda County Fair's exhibit supervisor, told me that feed costs have climbed steadily over the past few years, in part because of the ethanol boom, which has driven up the price of corn. The rising costs, combined with the recent recession, have resulted in fewer livestock projects, especially when it comes to steers, which require more feed than any other show animal. In 2006, the fair had one hundred steer projects; in 2011 it had just fifty.

And for owners of animals that don't go to auction—such as dairy goats like Chloe's and Serena's or pygmy goats like Allison's—there's no return on investment. Breeders generally charge between $150 and $350 for pygmy and dairy goats, which accumulate feed, hay, medicine, and vet bills over their ten- to twelve-year life spans. Both the Montclair and Pleasant Hill goat groups share the cost of supplies among the members. Allison estimates that she and her mom spent $100 on feed for goats last year.

All show animals also require accessories. Most showing products aren't expensive on their own—a half-gallon bottle of livestock shampoo might run you $20. But the full arsenal of supplies adds up quickly, and for the companies that make these products, 4-H is big business. During the months leading up to the fair, many feed stores hold clinics, where an expert comes to the store to give kids pointers on showing. Sometimes the experts are representatives from companies, and their tips consist of little more than instructions on how to use the product that they're selling.

I'm curious about the clinic phenomenon, so when Anthony tells me that a store called Nasco will be holding a "show and fitting field day," I jump at the chance to observe it. One Sunday, my friend and I make the drive an hour and a half east to the store's Modesto location. As we pull into the parking lot, we see that the place is packed with kids and parents. In addition to the clinic, the store is offering a 10 percent discount on showing supplies. By the store's entrance, we notice a large crowd around a booth with a sign that says "Glamb Jams." I have seen this product before: Kelly's mom, Teri, calls them "lambie jammies"—pj's for sheep, to keep the animals clean while they wait for their turn in the ring. The Glamb Jams are available in every imaginable pattern: hearts, pink camo, skulls and crossbones. The booth is decorated with photos of shorn show lambs sporting these fine garments. Also available are headbands for humans made of the same material as the jammies, should you want to match your sheep.

We follow the signs to a large yard in back of the store where the clinics are being held and sneak into a steer session that is already under way. Kids and their parents are paying rapt attention to an energetic young woman demonstrating the proper use of a conditioning spray. We stay for a few minutes, then wander over to another tent where someone is offering advice on how to use electric clippers to sheer a sheep.

Back inside the store a cavy breeder has set up shop in a back room. (One of the mysteries of 4-H is that members refer to guinea pigs exclusively as "cavies," ostensibly after the animal's genus, *Cavia*.) We watch incredulously as a potential customer scrutinizes a little nose-twitching cavy while a woman, presumably the breeder, holds forth about this cavy's exceptional great-grandfather.

After the clinics, customers roam the aisles and stock up on supplies: clippers, combs, and several different brands and formulas of shine-enhancing sprays for various species. It's like Sephora for show animals. Few customers leave the register with bills under $200. The Nasco catalog contains even more show accessories, including fake hair for cows ($49.25 for white or, puzzlingly, $109.50 for black), cow hair dryers the size of small vacuums in shades of metallic blue, purple, and green ($257–$353), and fetching Novelty Farm Socks for humans ($7.95).

Later, I called up Chuck Miller, Nasco's Wisconsin-based director of agriculture sales. Kids from 4-H and Future Farmers of America "are likely to be our future customers, so we want to find a way to engage with them, so that they know Nasco," he said. "They are a huge customer base for us." Nasco, he explained, is mostly a mail-order business. In addition to the showing and grooming catalog, the company offers a host of other catalogs aimed not only at farmers and ranchers but also at crafters, teachers, and others. Still, "livestock showing and grooming supplies account for a significant portion of the company's business—about 10 to 15 percent of Nasco's farm catalog total sales," he estimated.

"The stories that I hear—people can't wait to get the Nasco catalog," said Miller. "It's something that these kids are connecting with. The effort that they put forth to go to fairs and show animals—it's a big job." He compared the process of shopping for show supplies to acquiring backpacking gear. As an outdoor-store junkie with a weakness for things like collapsible camping wine glasses and ultralight inflatable pillows, I could relate.

The fact that not everyone has the resources to buy champion animals, expensive feed, and showing accessories is not lost on the National 4-H Council. In a section on ethics, an official beef curriculum offers a hypothetical scenario about two 4-H'ers, Billy and Bobby, who have decided to raise goats.[1] "Billy's mom tells him he can have anything he wants in order to take care of his goats, money is no problem," it says. However, Bobby and his parents "live in an apartment, and renting barn space is expensive," so "Bobby can only have limited funds." The curriculum then asks 4-H'ers, "How does the issue of money relate to show ethics? Would the competition be more fair if a spending limit was set or if there was no spending limit?" But the curriculum doesn't offer any solutions.

Sarah Watkins, California's 4-H program representative for animal science education at the University of California's Division of Agriculture and Natural Resources, notes that some 4-H clubs have tried to make showing more equitable by setting up monetary awards for some members whose families can't afford all the associated expenses. Still, Anthony's pig-group leader, Sally, finds it frustrating that the playing field isn't always even for the members of her pig project. In past years, some members of the group have chosen to buy inexpensive pigs and feed—but none of them placed in their market classes, and two were even disqualified for not weighing enough.

"With all the supplements and feeds available to make them appear better grown than they are, it becomes a competition to see which kids now have the right formula to get their pig the win," says Sally. "What does that teach kids? That you can buy success?"

Ranch kids often have a leg up, since for them, acquiring an animal with top-notch genetics is relatively easy: they can just pick one out of their family's stock. But none of the kids I had met so far lived on a ranch with a breeding program—or had the means or desire to spend thousands of dollars on an animal to take to the fair. I was beginning to see why many urban and suburban 4-H'ers focused on showmanship competitions—where the emphasis is on the kid's showing chops rather than the animal being shown.

CHAPTER FIVE

The Big Business of 4-H

The Nasco clinic that I visited opened my eyes to the perspective of the companies that sell showing supplies: they have correctly surmised that 4-H'ers are a valuable market—as showers of animals now, and even more so as farmers in the future. That kind of thinking isn't new. The connection between 4-H and the business world has deep roots.

In part, 4-H is a federal government program, housed in the USDA's National Institute of Food and Agriculture. Every year, NIFA earmarks funds—in 2012 it was about $341 million—for Extension activities, including many of the programs that fund 4-H initiatives at land-grant universities and state and county 4-H offices. How that federal money gets used is up to administrators at each Extension office, so the amount of USDA funding that 4-H programs receive varies widely from place to place. In addition to federal money, 4-H programs also receive money from state, county, and city budgets.

But when it comes to national initiatives, programming, and curricula, 4-H has another source of income: the National 4-H Council, the nonprofit that exists to fundraise for and promote 4-H. The council, whose 2013 revenue totaled about $40 million, has many different sources of income, including the 4-H Mall, an online store that sells products bearing the 4-H clover emblem; the National 4-H Youth Conference Center, which it rents out to groups; and investment returns. But today, the bulk of the council's funding comes from donations: from individuals, foundations, and corporate sponsors.

Almost since its beginning, 4-H has relied on support from businesses. In 1919, in Kansas City, Missouri, club leaders from across the nation met for the first time to compare techniques and formalize the club system. One of the attendees, a businessman named Guy L. Noble, convinced his employer, Armour & Company of Chicago, that it would be a smart business move to sponsor a national competition for members of pig clubs. Two years later, Noble temporarily took leave from his job to help organize the National Committee on Boys and Girls Club Work, a group that he hoped would raise money for the clubs. Noble called a meeting in Chicago, inviting other local businesspeople to start fundraising for 4-H.[1]

By 1923, Noble and his colleagues had secured donations from several large businesses. He enticed the funders with wholesome stories of youth transformed by the influence of the club; according to Thomas and Marilyn Wessel's *4-H: An American Idea,* at one point he brought two teenage dairy club members to the meeting of the American Bankers' Association in Rye, New York. The bankers were so impressed with the boys' demonstration of herd improvement that they named Noble's committee one of their top charitable causes.[2]

Another one of Noble's major projects was the club's monthly newsletter. In addition to reporting on 4-H events across the country, it published photos, contest announcements and results, and a section called "Sociability Lane" that proposed various ways for 4-H'ers to amuse themselves at meetings. (I have found that these ideas vary in quality. One low point was a 1939 newsletter that suggested a game called Guess My Weight, which is exactly what it sounds like. "The ladies especially will be curious to know how stout they appear," the article noted.[3])

In 1935, well into the Great Depression, Noble caused a stir when he opened up the 4-H newsletter to advertisers. Major accounts included Firestone tires, Electrolux iceboxes, Quaker Oats, and Montgomery Ward department stores. Extension leaders began to object: why, they asked,

NO. K-2. 4-H HATS

Made of double thickness of crepe paper, sewed. Suitable for either boys or girls. Makes fine appearance in a parade or at a 4-H club banquet.

10 for $ .90
25 for   2.10
50 for   4.00

100 to 500 for
$7.65 per 100

500 or more for
$7.50 per 100

No. K2

NO. 191. 4-H HATS

Unique design, attractive for girls. May be used in stunts, 4-H club parades, parties, banquets or on achievement days.

Same price as No. K-2.

No. 191

Figure 9. One source of income for 4-H was its supply catalog, from which members could buy fetching uniform hats like these. Gear emblazoned with the clover logo still draws revenue for 4-H today. From *4-H Handy Book* (Chicago: 4-H National Committee, 1928), 54.

should companies be allowed to make money off children? Noble angered them further by allowing advertisers to attend the yearly youth meeting in Chicago to display their products.[4]

In 1939, the Extension put an end to Noble's free-and-easy relationship with private sponsors, outlawing any sponsorship agreement that included advertising and declaring that the committee's newsletter was not even an official 4-H publication. Not one to be put off by a bunch of bureaucrats' rules, Noble continued to produce the newsletter as a private publication—with a robust circulation—for many years.

Interestingly, the Extension didn't put a stop to another one of Noble's fundraising schemes: sponsored contests. In the early 1920s, Noble had the idea of asking corporations to provide the prizes for 4-H contests. It worked. In 1922, the department store Montgomery Ward doled out trips to the annual national conference for club members with the best records in home economics projects. That same year, Atlas Glass Company sent winners of canning contests to France.[5] The model of corporate-sponsored prizes was extremely successful; the 4-H History Preservation Program lists 223 companies that donated awards over seven decades. The diverse group includes Campbell Soup Company, Chevron Chemical Company, Dr. Scholl Foundation, Ford Motor Company, Nabisco Brands, Pfizer Inc., Whirlpool Corporation, Wrangler Jeans, and Dow Chemical's Ziploc bags division.[6]

With his aggressive pursuit of sponsors, Noble, who didn't retire until 1958, laid the groundwork for a long and fruitful relationship between 4-H and the companies that he had courted.[7] As national contests declined in popularity, companies continued their support of the organization through corporate giving programs. In 1983, 4-H launched an ambitious new fundraising campaign, drawing $50.6 million in corporate donations over five years.[8] By 1988, its roster of supporters had grown considerably and now included the Coca-Cola Company, Eli Lilly and Company, Exxon Corporation, Fleischmann's Yeast, General Foods Fund, Gerber, Kellogg, Land O'Lakes, National Pork Producers Council, Orville Redenbacher's Gourmet Popping Corn, Pioneer Hi-Bred International, Purina Mills, and Singer Sewing Company, among many others.[9]

Over the next decades, 4-H's donors served the group well, supporting many initiatives. But in recent years, the National 4-H Council has courted businesses, foundations, and individuals more successfully than ever before. Much of that success has come out of what is likely the council's most ambitious project to date.

In the early 2000s, 4-H leaders took notice of reports that American students were rapidly falling behind their peers in science, math, and technology. When I spoke to Jim Kahler, the national program leader for 4-H science at the USDA, he remembered thinking when the reports came out that 4-H was perfectly positioned to nurture the next generation of scientists. "We had one hundred years of experience in science education," Kahler told me. "We knew we could do this right."

Indeed, 4-H has been honing its educational technique since the early days, when country kids tested various corn-growing methods to find out which one would work best on their farms. "We have always been about 'learn by doing,'" said Kahler, "and we know that when it comes to getting kids excited about science, this method works exceptionally well."

Judging from the 4-H'ers that I've gotten to know, I think Kahler and 4-H are on to something. When the only 4-H'ers I knew were Chloe and Serena, I suspected that they were good at showmanship and Rabbit Bowl because, with their 4.0 GPAs, they were clearly good at studying. After all, preparing for a rabbit judge's grilling essentially requires the same skills as cramming for a test.

But that theory didn't hold up once I met Allison and Anthony. Academics were not their thing; Allison actively disliked school. And yet they still devoured animal science for 4-H—and their reams of ribbons suggested that they totally killed it at the fair. Something about 4-H's hands-on method was helping them succeed where school had failed.

However, despite the club's hundred-year history with the learn-by-doing method, little formal research had been conducted on its efficacy. In

2004, national 4-H leadership began to evaluate 4-H programs to determine how successful they were in teaching science. The results were mixed. While science-based activities tended to generate a lot of enthusiasm among participants and leaders, they weren't happening everywhere. Club-based programs—especially those in urban areas—usually emphasized community service over science. "We noticed that some leaders were afraid to teach science, because they thought they weren't qualified," said Kahler.

Kahler and his colleagues knew that to bring learn-by-doing science to all seven million 4-H'ers would require a large-scale effort. They'd need new curricula and training materials, not to mention significant fundraising. So as 4-H leaders began to develop their science initiative, the council pitched the program to potential corporate funders as a long-term investment: by training children in science, technology, engineering, and math (STEM), 4-H was ensuring that its sponsors would have a workforce.

In 2008, with the help of corporate sponsors and a $750,000 grant from the math-and-science-focused Noyce Foundation, 4-H National Headquarters and the National 4-H Council partnered with the nation's 111 land-grant universities to launch the Science, Engineering, and Technology Initiative, later renamed simply 4-H Science. A team of veteran teachers was assembled to overhaul the science curricula used in clubs, camps, and after-school programs. Through the universities, 4-H set up intensive trainings for 4-H volunteers and leaders to teach them how to teach science.

But 4-H Science's most ambitious goal was to expand the organization's reach dramatically. Leaders pledged to attract a million new 4-H Science students by 2013. They knew that in order to do that, they would have to market the program effectively. This would be especially difficult because of a PR problem: the public still thought of the organization as just for the children of farmers—something that kids did between their barn chores. As Kahler put it, "'I'm not driving a tractor so why would I be involved in 4-H?' We want to fight against that assumption."

So the council's marketing team partnered with 4-H National Headquarters at the USDA and the Cooperative Extension to stage a major campaign to change the public's perception of 4-H. Most notably, in 2008, 4-H rolled out National Youth Science Day, a program that allows millions of students from all over the country to participate simultaneously in an experiment. A suite of new science programs followed. In 2009, the National Council launched 4-H Robotics, a series of curricula that allows kids in clubs, camps, and after-school programs to build real robots out of household materials. In 2010 4-H introduced Project Butterfly WINGS: Winning Investigative Network for Great Science, a program that teaches nine- to thirteen-year-olds across the country about butterfly ecology—and encourages them to enter their findings into a database that professional scientists use to track the health of ecosystems.[10]

Many of the most interesting initiatives happened on the local level.[11] In Bucks County, Pennsylvania, 4-H leaders developed a veterinary science program that teaches teenagers college-level anatomy and animal science through a series of hands-on clinics and field trips. At Rutgers University in New Jersey, a diverse group of fourteen- to sixteen-year-olds from around the state attended a weeklong camp that offered sessions on a broad range of science topics, including biomedical engineering, ocean science, and food science. At On the Wild Side, a weekend-long residential camp in California's Sierra foothills, fourth through sixth graders from low-income neighborhoods in Sacramento rotate through a series of experiments and hands-on projects that teach environmental science.[12]

"When we talk to firms that have need for STEM workers, they're ready to listen," said Andy Ferrin, an advertising industry veteran who has led 4-H's marketing team since 2008. "Leadership here has been very effective in reaching out to those that believe in our mission. We have found a lot of believers."

Indeed, the 4-H marketing strategy has worked. In 2007, $9.7 million of the National 4-H Council's funding—36 percent of its overall revenue—came from grants and contributions. Six years later, in 2013, that number had more than doubled; at $23.1 million, grants and contributions accounted for 58 percent of the revenue.[13] The list of corporations that currently sponsor 4-H's science programs includes, among others, Altria, AT&T, Cargill, John Deere, Lockheed Martin, Motorola, and Samsung.

When I asked representatives from the corporations that sponsor 4-H why they decided to support the organization, almost all characterized their donation as an investment in the future. "At a time when there is a high demand for skilled engineers and hard scientists in America's workforce, an increasing number of college graduates are opting out of careers in the needed technical fields," Kirsten Norman, a spokeswoman from the agribusiness company Cargill, wrote to me in an e-mail. Because of this growing deficit of skilled workers, she continued, Cargill supports 4-H's science and math programs.

Matt Blakely, the head of Motorola's charitable foundation, echoed those sentiments. "As a company of engineers and scientists and technology professionals, we think it's important for our company to support the next generation of professionals in those fields," he told me on the phone in January 2013. Since 2007, Motorola has given $550,000 to 4-H's STEM programs. Blakely explained that Motorola chose to support 4-H because of its impressive diversity. "They have established relationships with rural areas and small towns," he said.

In 2011, five years after developing its science initiative, 4-H headquarters released a study about the program's impact, surveying fourth, eighth, and twelfth graders.[14] Its findings were encouraging:

+ Of fourth graders who had been exposed to 4-H programs, 55 percent thought they were good at science, compared to 42 percent of their peers.

+ While just 40 percent of fourth graders nationwide thought that science was useful in solving everyday problems, 61 percent of 4-H fourth graders did.

- Among eighth graders, nearly a third overall thought science was boring, compared to just 14 percent of 4-H'ers.

- In twelfth grade, 73 percent of 4-H'ers listed science among their favorite subjects; 40 percent of their peers did.

- Of twelfth-grade 4-H'ers, 67 percent said they wanted to pursue a career in science, compared to just 37 percent of their peers.

Another report found that tenth through twelfth graders in 4-H were twice as likely as their peers to take part in science programs outside school.[15] In twelfth grade, 4-H girls were three times more likely than their female peers to participate in science programs.

In 2012, when 4-H surveyed county Extension agents to find out whether they had noticed a difference in science education in their communities, three-quarters of the agents reported seeing an increased emphasis on science since 2006.[16] Because 4-H doesn't administer the same kinds of standardized tests as public schools, we don't know as much about how much the students learn. But what we do know is promising. By 2013, 4-H had exceeded its goal of attracting a million new students to its science programs.

The more I learned about 4-H's science education program, the more impressed I became. I loved that 4-H had succeeded in engaging kids—especially girls—in the very subjects that most of us learn to think of as difficult and boring. American schools, I thought, could learn a thing or two from 4-H's approach.

And yet the corporate sponsorship element made me a little wary, for the same reason that Guy Noble's 4-H newsletter advertising program must have worried the Extension leaders. I came across an online fundraising tool kit, published by the National 4-H Council and a management consulting firm called the Osborne Group, that instructs local 4-H leaders on how to raise money to support their programs.[17] The guide gives several examples of how companies have used 4-H's reach and reputation to their own advantage.[18] A section called "Providing Business Value" recounts how J.C. Penney wanted "access to customers" and "increased brand loyalty"; 4-H developed a partnership wherein the department store chain funded 4-H after-school programs in poor communities. The programs, which were all set up within twenty miles of J.C. Penney stores, also provided J.C. Penney gift cards to needy families.[19] Similarly, the farm goods store Tractor Supply Co. sought "access to new customers"; 4-H says it "increased store traffic" through a program where customers could buy a paper clover to display in the store—with proceeds going to local 4-H clubs and the National 4-H Council. To the council and the corporate funders, arrangements like these must have looked like win-win situations. But I didn't love the idea that 4-H—which also received my taxpayer dollars—was drumming up more business for companies.

Then, I came across the website of the 4-H AgriScience program, meant "to cultivate the emerging study of biotechnology and business/economics in the agriculture industry through hands-on experiential learning activities and online learning courses for youth."[20] One of the program's main sponsors was the global chemical and agriculture firm DuPont, a Fortune 100 company. DuPont executive vice president James C. Borel is the chair of the National 4-H Council's board of trustees.

From the AgriScience website, I clicked through to several different curricula. The facilitators' guide, which is sponsored by the United Soybean Board, says that the links included "have been reviewed and are reliable sources of additional information," and the fine print on these lesson plans notes that "no endorsement of a commercial entity or its products or services is intended or implied."

But at times, it looked to me like 4-H'ers were indeed being given a dose of advertising along with their educational material. For example, students learn that the answer to the title of the lesson "What Makes NesQuik™

Quick?" is the surfactant soy lecithin—of which DuPont is a major producer. The lesson ends by instructing students to "follow the directions on the label and make yourself a snack drink with NesQuik™."[21] In sidebars scattered throughout the lessons, students are encouraged to click on links that lead to supplemental material online. Some of the links take students to DuPont's website. In the "news and careers" section of a lesson called "Face the Fat: Engineering a Better Oil," students who click on "Plenish™ Is a New Oil" will find themselves on a DuPont web page promoting a new soybean oil that offers "an attractive trans fat solution to the food industry."[22] A link in the lesson "Agriculture at Work: Bioplastic" takes students to a press release about DuPont's formation of an industry group called the Starch-Based Biomaterials Alliance.[23]

Although DuPont no longer funds the AgriScience program, its curricula are still available on the website. When I asked Michelle Gowdy, director of community and academic relations at DuPont and president of the Pioneer Hi-Bred Foundation, why DuPont sponsored the curriculum, she explained that Pioneer Hi-Bred, DuPont's hybrid seed division, was planning to increase its workforce dramatically. Since 2009, the company has hired a thousand new employees annually, and it plans to continue this trend over the next few years. "With United States science education stuck in the doldrums, where will those employees come from?" she said. "We need more people interested in careers in agriscience."

In addition to the curricula available online, 4-H AgriScience also includes a program called Teens Teaching Youth AgriScience/Biotechnology. Sponsored by the United Soybean Board, the program targets kids in cities. According to a 4-H evaluation, when students were surveyed about the program's most valuable lessons, they reported having learned that "soybeans are a huge part of many things used daily," that there are "lots of alternatives to crude oil," that "bio-plastics are eco-friendly," and that one "can use soybeans to make common products."[24]

All of that is true, of course. But I wondered whether students in 4-H's AgriScience programs were hearing the whole story—learning about the drawbacks of industrial agriculture as well as the benefits. While I saw plenty of good news about the power of biofuels to replace petroleum in the AgriScience curricula, I didn't see a single mention of the effect of converting vast swaths of farmland to fields for corn to make ethanol—how soil scientists have warned that planting so much corn year after year will eventually strip the land of its valuable nutrients. Although the AgriScience/Biotechnology Facilitators Guide does include a lesson about the controversy around genetically modified organisms, it doesn't once mention the growing body of evidence that crops engineered to withstand herbicides require an ever-increasing amount of chemicals—because weeds evolve to resist the herbicides, as well.[25]

When I asked Andy Ferrin, 4-H's chief marketing officer, whether he was concerned about the implications of allowing agribusiness firms to sponsor a curriculum, he told me that he didn't believe that the curriculum was one-sided, and the National 4-H Council said that sponsors did not contribute to its content. Besides, Ferrin said, 4-H has a long tradition of exposing children to cutting-edge agriculture technology. "Of course 4-H has always been about science and technology," he said. "Young people convincing their families to change the way they operate their farms to produce more food. It has been science in action from the very beginning."

At the same time that 4-H has been tapping corporations to support its science programs in the United States, the group has made a major push to ramp up its presence in the developing world—and there, too, it has cozied up to agribusiness to support its work. In 2011, 4-H launched an initiative to promote youth development and improve food security in rural farming communities in Africa.[26] On the surface, this program closely resembles the model of the earliest 4-H clubs in the United States: Leaders help kids plant gardens. The children learn new farming techniques so that they can increase their yields and sell the produce in the community while also teaching their parents the new methods. The program, funded in part by the Bill and Melinda Gates Foundation, has been hailed by both American and African leaders as a success; some clubs have made enough money from sales of their crops to help needy students afford school fees, uniforms, and other supplies. The adult farmers in the villages have benefited, too, by using the new techniques to make their own farms more productive.

But the program has another, more controversial side: One of its main sponsors is DuPont's Pioneer subsidiary, one of the biggest players in the global seed industry. For the first year of the program, Pioneer supplied all the hybrid maize seeds for the 4-H gardens free of charge. The sponsorship, DuPont says, is part of a multimillion-dollar initiative to promote food security in the developing world. As DuPont's food and agriculture outreach leader, Lori Captain, put it to me, "This next generation of youth is really going to be critical to addressing the food security issue, the nine billion people that we will have on the planet by 2050. We need to engage youth all over the world to help address food security. We want to build the skill and build the will."

DuPont claims that it is doing a service to farmers in Africa by introducing its nutritious and high-yielding hybrid. But there's one major drawback: DuPont's seed costs significantly more than the local variety—as much as ten times more, according to some of the agriculture leaders I talked to. What's more, because it is a hybrid, its seeds can't be collected and used again—if farmers try that, the crops return to their parent varieties in just one or two generations.

Still, DuPont has found a willing partner in 4-H. "We want eleven- and twelve-year-olds to gain an appreciation for agriculture as a possible career and a business opportunity," said Shingi Nyamwanza, the managing director of 4-H's programs in Africa. "DuPont is making that possible." In 2012,

3,500 children between the ages of six and nineteen enrolled in the program in Ghana. In 2013, 4-H and DuPont started similar school garden projects in Tanzania and Kenya. In 2014, the program will expand to Ethiopia and South Africa.

The Africa program developed by 4-H is part of a larger trend—the continent's transition from subsistence farming to industrial crop production. American agribusiness companies have found a vast new market for seeds, fertilizer, and other farming inputs in Africa. Indeed, the 4-H gardens "are essentially demonstration plots," DuPont's Captain told me. "We are going to work with you, and you can show your neighbors in your tribe and village what you can do. And then they will want the seed as well."

But to other people, it looks more like DuPont is using 4-H kids and their gardens as free advertising for its new product. They warn that the seeds are too expensive for subsistence farmers to buy, and to grow: high yields typically require more chemical inputs—like pesticides, herbicides, and synthetic fertilizer—than the old seeds. Devlin Kuyek, a senior researcher at the global sustainable farming advocacy group GRAIN, told me he has seen this dynamic play out before. "They're giving samples to these kids to try to get Africans to accept them and use them," he said. "If you get farmers to switch to that and give up your local seed, then the private seed companies can make money."

CHAPTER SIX

"We Are Praying That DuPont Will Continue to Provide for Us"

Reader, I went to Ghana. Now, you may be thinking: what a strange place to look for authentic 4-H'ers. And it's true—had I wanted to find more of them, I probably should have headed somewhere like Iowa or Nebraska, not Africa.

But to tell you the truth, the more actual 4-H'ers I met, the harder it became to conjure that original wholesome youth of my mind's eye. I wondered whether a 4-H'er in Ghana would have anything at all in common with Chloe, Serena, Allison, Anthony, Randy, or Kelly. I knew that the 4-H program in Ghana was very different from that in California; the focus was on growing maize and vegetables, not on raising animals. Still, I hoped that maybe there was some essential 4-H quality that members of the two groups might have in common.

And what I had learned about 4-H's new initiative in rural Africa was too intriguing to pass up. I read that 4-H had programs in more than fifty countries, and that many of these international programs had been around since before World War II. Most of these programs were self-contained and, aside from exchange programs, had little interaction with American 4-H headquarters. But a Duke University historian named Gabriel Rosenberg told me about one fascinating example of U.S. 4-H's involvement with clubs abroad.

In his forthcoming book, Rosenberg chronicles how in 1960, Howard Law, an employee of Nelson A. Rockefeller's American International Association for Economic and Social Development, took over a network of 4-H clubs in Latin America, securing loans for them from Bank of America, Chase Manhattan, and the Inter-American Development Bank.[1] The program also received funds from the United States Agency for International Development (USAID) and some of the largest agribusiness companies at the time—including Eli Lilly, Purina, and General Mills. In addition to arranging for loans, Law and his team taught the clubs new farming techniques and provided them with high-yield hybrid corn and fertilizer.[2]

In 1968, 4-H leaders in the United States took over Law's program and ran it until 1975, when it finally dissolved. Although other 4-H programs abroad continued, they did so mostly without the support of American 4-H. I found it intriguing that the National 4-H Council today seems to have a renewed interest in its programs in the developing world—once again with the support of American business.

So I boarded a flight to New York's Kennedy Airport and then picked up a nine-hour connecting flight to Accra, the capital of Ghana. After a bleary night in a hotel there, I traveled north to Koforidua (Ko-fo-RHI-jyah), the small city where 4-H Ghana has its headquarters.

My first morning in Koforidua, I am sitting in a packed-to-the-brim beater of a non-air-conditioned van in the middle of a dusty, unpaved parking lot. "Jesus is the savior!" yells a man dressed in a suit leaning into the van. Beads of sweat cover his bald head, and his voice is hoarse from yelling about God. "Eggs! Eggs for sale!" calls a woman carrying a tray of hard-boiled eggs on her head. "Mentos!" announces another guy as he reaches through the van's open window, tugs on my sleeve, and gestures to the packages of mints he carries in a basket. The air is filled with fumes from dozens of other vans—

called *trotros* here in Ghana—and even though it's before 9 A.M., the heat is intense. I guzzle bottled water.

The preaching guy leans in further, so he's yelling only a few inches from my ear. He is reaching some kind of crescendo, but no one seems to be paying attention. I turn to my traveling companion, Steve Koranteng, a 4-H project assistant, who is sitting in the row of seats behind me.

"Isn't everyone here Christian already?" I ask him. "What is this guy doing?" In my first few days in the country I have noticed that evangelical Christianity saturates Ghana. In the hostel where I am staying I am awakened every morning at five by the sound of a man yammering about the Lord in the hallway. The hotels that we pass in the city bear banners advertising visiting preachers, and the roadside stalls have names like Jesus Has Made It Tilapia or God Is My King Hairdresser.

Steve shrugs. "He is moved by the spirit."

"God is the only one!" roars the preacher. "Accra, Accra, Accra!" calls the driver of a *trotro* headed to the capital city. "Obroni!" says another hawker, using the word for "white person" in the local Twi language and squeezing past the preacher to try to catch my attention. I take another swig of water as two more passengers get in. The van is now full, so the driver slides the door shut and takes off, barreling down a partly paved road, heading up into the hills outside Koforidua. Steve is bringing me to the Ehiamankyene (Ehi-a-MAN-chih-nay) School, a two-hour drive north. Like most of the 4-H clubs in Ghana, Ehiamankyene's club has both a vegetable garden and a separate plot devoted to maize—a very important crop in Ghana. It is the basis for one of the country's staples, a fermented dough called *banku* that Ghanaians like to dip in soup.

Most of the students at Ehiamankyene come from subsistence farming families. Although the majority have enough to eat, Steve tells me, few of the families have any extra money, let alone savings accounts or credit cards. Most of them don't have refrigerators and rely on outdoor fires to cook their food.

After a long, bumpy ride, we arrive at the school, a collection of several simple cement structures packed with students in brown-and-orange uniforms. The school's principal leads us to the classroom where the members of the 4-H club—mostly middle schoolers—have assembled. Steve introduces me to the club leader, a twenty-six-year-old Ghanaian science teacher named Charity, who faces the group and claps her hands.

"Baah!" she says. "Where is Baah?"

A tall, thin boy stands. He smiles, shakes my hand, introduces himself as Francis Baah, the club's president, and offers to show me his club's maize plot. I follow him and a few of the other club members down a red dirt road with palm jungle on either side.

Francis and I chat about school, which he loves. He lights up when he talks about farming and his dream of studying agriculture at university one day. At eighteen, he is only in the equivalent of eighth grade, so he still has four more years of school to go before he graduates. Like many kids in Ghana, Francis attends school only when there's enough time and money.

Francis hasn't always wanted to be a farmer. As a child growing up in this rural village, he watched his father toil all day in his fields of maize and cassava. His mother would make bundle after bundle of dried maize to sell at market. And even after all that hard work, there wasn't always enough money to send Francis and his four siblings to school.

In the years when the crops were good and Francis's family did have enough money for school, he was an excellent student, always at the top of his class. He planned to be a lawyer or an engineer and maybe someday move to Accra, where there are more jobs.

But then, when Francis Baah was sixteen, all that changed. One day, he heard that a teacher at his school was going to start a new agricultural club called 4-H. Even though Francis wasn't interested in agriculture, he made a point of participating in as many school clubs as possible, so he decided to

"Praying That DuPont Will Provide"

join. To his surprise, he found that he loved it. Visitors from the 4-H office in Koforidua taught the students all about better methods of farming—planting maize seeds in rows instead of simply scattering them, for example—and introduced them to a special new kind of maize seed called Pioneer. They helped the club members start a vegetable garden and a maize plot and gave them the special seed for free.

That year there was very little rain, and Francis's father's maize crop was small. But the 4-H club's maize thrived, and when Francis tasted it he couldn't believe how delicious it was: sweeter than the maize that his parents grew, with fewer fibers to get stuck between his teeth.

Francis's 4-H club sold its harvest at the local market. With the proceeds they bought a new fence for their vegetable garden. Each 4-H member was also given some of the money to take home. The club leaders from the city told them that with the new techniques and seeds, it was possible to make a good living as a farmer. Francis went home and told his parents that that's what he wanted to do. At first they were disappointed—he had the potential to do something better, they thought. "But then I educated them about being a farmer," says Francis. "I told them that I could have a farm with a tractor and hire people to work for me."

Francis showed his parents all the money he had made from the 4-H gardens. He brought them to see the plots. "My family thought our land could produce only maize," he says. "But after 4-H, they also started cultivating vegetable crops." Francis's dad was especially interested in the new maize seed that his son's club was using, so Francis asked the 4-H club leader if he could have some Pioneer for his dad to try. The next time a representative came from the 4-H office, he brought extra seed. Francis's dad tried it, and he found that it required less water than the old kind, and it also yielded more ears.

Francis leads Steve and me off the road down a dirt path in the jungle to a one-acre plot with maize sown in rows. The plants are about as high as my waist. Since they are hybrid plants, the club won't be able to collect the seeds from this crop to use for next harvest, the way that many subsistence farmers have done for ages.

"So how is your club going to get more seeds next year?" I ask.

"It depends on the 4-H office, whether they will give it to us or not," says Francis with a grin.

Next we set off to see Francis's family's farm. He shows Steve and me the kitchen—a fire ring, a few large pots, and some tools under a piece of cloth supported by wooden posts—and then we walk over to Francis's goat and sheep enclosures, roomy cages made of branches that Francis has gathered from the forest. Inside, a few animals jump around and bleat insistently, excited to see Francis. One of the goats—which looks similar to the Nigerian Dwarf breed that Chloe and Serena raise—is only a few days old. Francis scoops her up and she looks indignant, wagging her tiny tail. Francis says that he feeds his goats and sheep a diet of palm fronds and cassava husks. Once the goats are mature, he sells them for meat.

Francis goes to find his father on the family's maize farm, on the other side of a patch of jungle. A few minutes later he returns with a friendly, white-haired man who looks to be in his forties or fifties. With Francis and Steve translating from Twi, Francis's dad tells me that he is very pleased with the sample of Pioneer seed that Francis's 4-H club has given him. This year, his plot was extremely productive, and the maize tasted better than the kind he used to use, called *obatanpa*. Developed by Ghanaian crop scientists in the early 1990s, *obatanpa*, which means "good nursing mother" in Twi, is high in protein and the essential amino acids lysine and tryptophan.

"Now that I have this knowledge about the Pioneer corn, it is the only maize that I want to plant," he tells me.

"So will you buy more?" I ask.

"I don't have enough money," he says. "That is the challenge right now."

Figure 10. Ghana 4-H member Francis Baah, 18, holds his baby goat inside the cage he made out of wood he gathered from the jungle. Photo by Kiera Butler.

On the van ride back to Koforidua, I ask Steve about the Pioneer maize. What's the likelihood that Francis Baah's father will be able to get more of it?

"Pioneer is not giving out any more seeds," says Steve. "Not to 4-H, and not to the farmers. We are really hoping that Pioneer will decide to give out more seeds, or that they will figure out a way to make the price lower. Otherwise, there is no way that farmers can afford it."

By the time we arrive in Koforidua it's evening. We head to the 4-H office, three small rooms crowded with desks overlooking a dusty vacant lot ringed by a crumbling cement wall. Next door is something called the Child Evangelism Fellowship, and downstairs is a little stall that sells water, soda, and packages of biscuits. Even though it's close to 7 P.M., all the 4-H employees are still in the office. Steve finishes up some work, and then we go out for dinner. Steve leads me through alleys and markets, stopping occasionally to greet friends. We sit down at an open-air restaurant with plastic tables and chairs—one of Steve's favorite spots.

Steve is twenty-six and has lived in Koforidua his whole life. He rents a modest apartment for about $35 a month, but he barely spends any time there: in addition to 4-H, he runs a call-in radio program for Ghanaian youth and works on his screenplay, a romantic comedy that he hopes to sell to some producers in Nigeria. His favorite movie is *Let It Shine,* a 2012 straight-to-TV Disney film with a Cyrano de Bergerac–type plot. He loves *banku,* and also hamburgers.

There are no hamburgers at this restaurant, so we order the *banku,* which arrives in baseball-sized mounds wrapped in plastic. Steve shows me how to make little balls of the dough to dip in the bowl of spicy, soupy sauce that comes with it. I try really hard to like *banku,* but unfortunately it reminds me—in both texture and flavor—of Play-Doh. Steve polishes off two *banku* mounds; I make it through about half of one.

"I think *banku* might be an acquired taste," I say. Steve laughs.

Over the course of the next few days Steve and I visit a half dozen other 4-H clubs. I have a blast talking to the kids, who are eager to show me their record books and brag about how much money their club has made. At one school, a quick fourteen-year-old named Adelaide answers all my questions before any of the boys in her class can get a word in edgewise. On our way back from her club's maize plot she makes sure to walk next to me, and she tells me she wants to be a farmer when she grows up. Then she stops, looks me in the eye, and asks if I can take her to the United States so that she can learn our farming methods and bring them back to help people make a lot of money. I don't know what to say, so I tell her I wish I could and that she should stay in 4-H because I think they do exchanges. She nods seriously

Figure 11. Members of a primary school 4-H club in Ghana show off their "garden eggs" (small eggplants). Photo by Kiera Butler.

and thanks me for coming, and I feel terrible because I made up the bit about the exchanges. I have no idea whether 4-H or anyone else will ever help Adelaide achieve her dream of learning American farming.

One of my favorite clubs is at a primary school up in the hills, where the air is a little cooler, in a village that is known for its long-lived residents. Steve claims that people here live until they are 120 or even 130. I tell him I'm skeptical. The 4-H'ers in the village, ten- to twelve-year-olds, crowd around me, excitedly showing me the "garden eggs" (small Ghanaian eggplants) that they've grown in their 4-H plot. One little boy asks me if I'm friends with the hip-hop star Chris Brown. Another asks, earnestly, whether President Obama is my husband.

"Do I look like Michelle Obama?" I retort.

"Yes," he says, completely straight-faced. He darts away before I have a chance to ask him whether he has noticed that I am a pale and freckled white person.

A girl points at the maize plot that she and her friends have just planted. I ask her whether her parents have seen the 4-H garden. She tells me they have. What do they think? I ask. She looks down and whispers, "Proud!"

Everywhere I go, the 4-H'ers and their club leaders rattle off the ways in which the DuPont corn is superior to the *obatanpa* maize. And then they tell me they can't afford the seed or the tools to scale up. Some of them ask me for donations.

Later, I call DuPont and ask a spokeswoman why the company is trying to sell maize that none of the local farmers can afford. A few days later, the woman forwards me an e-mail from Worede Woldemariam, the company's senior business manager of operations in Africa. "DuPont Pioneer's pricing strategy is a sustainable, balanced approach that sets prices based on a number of factors," the e-mail states. "The responses we are seeing from existing Pioneer customers in Ghana is proof that our leading technology, superior agronomic performance and unmatched customer service is very much appreciated by our customers." I e-mail 4-H to ask why the club is using seeds that are too expensive for most farmers. Shingi Nyamwanza, who coordinates 4-H's programs in Africa, responds that the seeds are "valuable for establishing the infrastructure of the program" and that farmers rave about them. "Beyond that feedback," she writes, "we are exploring the relative benefit of the seed compared to the higher cost."

One day during my trip, an agricultural ministry representative named Francis Nii Clottey stops by the 4-H office in Koforidua. He declares Ghana's 4-H gardens a success: The kids have learned that farming can be fun and rewarding, and in many cases they've been able to supplement their diet with fresh fruits and vegetables. They've taught their parents tricks to improve their own home gardens. He estimates that two hundred farmers in Ghana have adopted the 4-H methods.

But he has reservations about the Pioneer maize. In his district, after a school 4-H club had a field day and showed off its new maize plot, fifty local farmers signed up to buy Pioneer seed—but in the end only six could afford it. Clottey and the other farming ministry representatives are trying to persuade the government to subsidize the cost of the maize, at least for a little while. But even if the government does step in, he says, he isn't wild about the idea of local farmers buying exclusively from an American corporation. "To some extent it worries me that we will have to rely on DuPont over and over," he acknowledges.

And besides, the farmers need so much more than just seeds to succeed. Equipment, transportation, fertilizer, and pesticides are also required and have to be bought year after year. Though the 4-H curriculum recommends using natural pesticides, Clottey says that method is not practical for larger-scale farmers.[3] "If you look at the natural pesticides, they break down very quickly," he explains. "So for commercial farmers, that doesn't work. It comes down to economics. You have costs involved when it comes to multiple applications."

After speaking to Clottey, I call AgriServe, the Ghanaian company that sells DuPont seeds. I ask Raja Najjar, the company's CEO, whether he plans to ask DuPont to lower the price of the maize seeds for the farmers who can't afford it. "We can't ask Pioneer to make this seed more affordable," he answers. "Once the farmer sees the yield, he will forget about the high price."

In other countries, that strategy hasn't always worked as planned. Pedro Sanchez, director of the Agriculture and Food Security Center at Columbia University's Earth Institute, told me about a similar program in Malawi. In 2006, the Malawian government introduced hybrid maize, offering farmers a 70 percent discount on both seeds and fertilizer, since the hybrid required a significant amount of synthetic fertilizer to thrive. In less than a decade, the country almost tripled its yields.[4] For the first time in recent history, farmers actually have a surplus of maize instead of a shortfall. "The farmers are well fed, well dressed," said Sanchez, who helped the Malawian government develop the program. "They are so much healthier now."

Those gains are real, but they came with a price. Sanchez and the other project organizers are trying to convince the farmers to use organic fertilizer instead of synthetic stuff, which requires significant amounts of natural gas and can pollute waterways. But the farmers "told us they only wanted the inorganic fertilizer because they want results right away," said Sanchez. Moreover, eight years into the program, Malawian farmers are still dependent on government subsidies for both seed and fertilizer—they haven't yet broken even.

One hot afternoon toward the end of my trip, as I am leaving a club's maize plot at a school in a small village, a science teacher stops me. He implores me to write an article that will convince DuPont that Ghanaian farmers need more free seeds. He explains that the farmers in his community can't pay full price. Someday, he says, they hope to increase their yields enough that they can afford Pioneer seeds—and all the land and inputs those seeds require—year after year. But in order to do that, the farmers need a boost. "Please tell DuPont to give us more seeds; we don't have wings to fly," he entreats me. "We are praying that DuPont will continue to provide for us."

That night, my last in Koforidua, the garrulous director of 4-H Ghana, Appiah Kwaku Boateng, known in the office as "Boat," invites Steve and me to his house for dinner. From the 4-H office, we drive in Boat's beat-up old car for a few minutes to a quiet, residential neighborhood at the edge of the city. We pull up to a small house, where Boat lives with his wife and their four children. I love Boat's house the minute I step inside. The living room walls are covered in a chaotic and cheerful bunch of family photos. A table and chairs occupy the middle of the small room, and an old comfy couch sits to one side. A TV on low volume hums in the corner. A kitten plays under a chair. Boat's kids tumble in and greet their dad in Twi. He jokes with them and proudly lists their accomplishments at school. He tells me he believes that if Ghanaians invest time and energy in children, the country's future will be brighter.

We sit down at the table, and Boat's wife brings out big bowls of food: chicken in a tangy sauce and a local dish called *jollof* rice, which is similar to a pilaf. We eat from paper plates inexplicably decorated with pictures of Thanksgiving turkeys, and we drink Coke from plastic cups. After the meal, Boat's wife, who looks exhausted from cooking dinner and working all day, naps on the couch while two of Boat's sons entertain me with "magic tricks" that involve a lot of giggling and tearing up newspaper. When that gets old, the youngest son brings out a family photo album, narrating as he turns the pages: "That's my mom. That's my mom's friend. That's my mom with me in her belly." There are a few photos of the family at church, and I learn that Boat and his family are Mormons, and that Boat serves as a bishop. I tell them that I was born in Utah, which they get a kick out of. They've never been, but they hope to visit someday. The kids and I both start yawning. It's late, and I have to begin my long trip home the next morning. Full of good food and happy to have made some new friends, I say goodbye to Boat and his family. We promise to keep in touch.

I leave Ghana feeling more confused than when I arrived. On the one hand, I am deeply impressed with Steve's and Boat's dedication to 4-H. I have watched them put in long hours at the office to make sure that their programs are reaching students in remote places, and that those students are coming away from the program with a positive impression of farming. They strongly believe that Ghana's quickly developing economy needs a new generation of farmers.

But I am not so sure about DuPont's role in the program. It seems to me that Ghana's 4-H'ers are learning that in order for their gardens to succeed, they will have to keep buying expensive seeds from DuPont—and chemicals from other corporations. How can young farmers learn to be self-sufficient if they're told they need to rely on American companies season after season? I can't help but wonder whether DuPont's main objective is less about helping farmers than about finding more customers.

CHAPTER SEVEN

Winning Champions Love Root Beer

So what did 4-H'ers in Ghana and California have in common? On the surface, not much. After all, it would be hard to find two places more different
than Francis Baah's small village and the tree-lined streets and strip
malls of the suburbs where many of my 4-H'ers lived. The stakes of 4-H
in the two places were different, too. While my California crew merrily
obsessed about trivial things like Rabbit Bowl and master showmanship
and plaques that said "champion wether dam," Francis Baah and his
classmates were focused on expensive new seeds that could change the
way their families farm. I suspected that most of my California 4-H'ers
would not grow up to become farmers and ranchers; the closest would likely
be Kelly Semonsen, who planned to put her sheep experience to work one
day as a livestock veterinarian. For Francis Baah, farming was a real career
possibility. The lessons that he was learning in 4-H could change the rest of
his life.

But in other ways, ambitious, focused Francis Baah was not so unlike
Kelly or any of the other California 4-H'ers. The Ghanaian 4-H'ers had
patiently shown me their thriving maize plots, explaining exactly how far
apart they had spaced the seeds and why it was important to weed regularly.
They were so proud of their work. It was the same pride that I had seen in
Serena, as she described how she delivered Huckleberry and Finn, and in
Allison, as she taught the younger Kayla how to care for her new goat. The

4-H'ers I had met in both the United States and Ghana had gotten a sense of what it feels like to be an expert at something. And they liked it.

By early April, my California 4-H'ers are beginning to prepare for the fair in earnest, attending showmanship practices after school and on weekend afternoons. One Monday evening I drive out to Martinez to see Anthony's pig group. The meeting is at the home of Sally Pereira-Cox, the group's leader. Sally, who is about my age, has wide blue eyes and wavy dark hair that frames her youthful face. Smart, upbeat, and organized, she has been involved with 4-H almost her whole life; Anthony's mother was her 4-H leader a decade ago. Now she works as a pastry chef at a restaurant in Napa and has two tow-headed stepsons, ages ten and eleven, who are members of her pig group.

When I arrive, I hear laughter and shouting coming from the backyard. I walk around the side of the house and see the members of the pig group rolling beach balls along the ground with pig sticks. There's a lot of giggling and, in one corner of the yard, an extracurricular game of tag. One of Sally's boys wallops his beach ball.

Sally makes her way over to me from across the yard. "They're supposed to be practicing showmanship," she says, rolling her eyes. "The beach ball is supposed to be the pig, so obviously they're not supposed to be hitting it that hard." A preteen girl shrieks gleefully as she slams her crop down on her ball, sending it flying. It deflates upon landing, and she howls with laughter. In a corner of the yard, I see Anthony calmly strolling after his ball, seemingly unperturbed by the mayhem around him.

Sally calls for everyone to stop. "Anthony, you're the judge, so you stand in the middle. Now everyone move your pigs around Anthony, like you're in the show ring." Anthony obliges and saunters into the middle of the group, turning every so often to see the kids around him. Is he bored? Proud that Sally has chosen him to play the judge? It's impossible to tell. His facial expression doesn't give anything away.

For a few minutes things seem more orderly, but then two of the boys start racing, and pretty soon everyone is back to yelling and punching the balls.

"That's it!" calls Sally. "Anthony, who's the winner?"

"Uh . . . Johnny, I guess," Anthony offers with a shrug. "Since he's the only one making eye contact with me." Johnny is an older member of the pig group, a good-looking high school boy who has been halfheartedly poking his beach ball and tolerating the younger kids.

"Okay, everyone," says Sally. "Who has a tip for showing? Johnny, you've done this a bunch of times before. What can you tell kids who are going to the fair for the first time?"

"Um," says Johnny. "You have to shave your pigs so the judge can see the muscle tone. Then you put oil on them."

Anthony raises his hand and holds up his plastic pig stick. "The judges like these now. They sell them at Concord Feed. Don't get the leather slappers, because the judges don't like that."

A high school girl volunteers, "The judges might ask you questions. You might have to say what ratio of protein you feed them. Also, you always want the pig between you and the judge. Act like you know what you're doing, even if you don't."

"Smile, smile, smile," says Sally. "Look as cute as possible. You need to read nonverbal cues from the judge, which is why it's important to maintain eye contact."

"If your pig runs away, you shouldn't run after it," chimes in one of the boys. "You should walk."

"If your pig is aggressive toward another pig, don't try to stop the fight," adds a girl with red hair. "Just back off, because you don't want to get hurt."

"When you're in the pen, never stop showing," says an older girl.

Sally thanks everyone for their suggestions, and then says, "If you're keeping your pig at the Wright barn, I want you to go downstairs, because Mr. Wright has a few things he wants to say to you."

Sally explains that Mr. Wright, a friend of her family who lets 4-H kids keep their pigs on his property for free, is not happy about the condition of the barn. Cleaning is mostly the responsibility of the 4-H'ers who board pigs there, but the whole group is supposed to help out—especially since they keep a collectively owned "restaurant pig" there. This pig will be slaughtered in two months for a 4-H pig-roast fundraiser held at the restaurant where Sally works.

After a few minutes the group from downstairs comes up, looking chastened. Sally decides to call it a night, and the meeting is adjourned. The next time the group will gather is to weigh their pigs a week before the fair.

Up in the Oakland hills, three new members have joined the Montclair 4-H goat herd. Kajsa has had her kids—a doeling (female baby goat) and two bucklings (male baby goats). Showmanship practice is scheduled for the day after they're born, so I head to the hills to meet the babies and see how everyone's showing skills are coming along.

Chloe takes me down to see the newborns, whom she has named after characters from the books that she is reading for her AP English literature class: Cyril from *A Passage to India*, Nerissa from *The Merchant of Venice*, and Gatsby from *The Great Gatsby*, all books that she loved.

The three babies are even tinier than Huckleberry and Finn were when I first met them. Cyril is the clear runt, and he hasn't learned how to nurse yet. Chloe and Andy are worried; they're considering trying to bottle-feed him if he doesn't figure it out today. The younger members of the goat group pass the new babies around, not altogether gently. But the tiny goats don't seem to notice. They look up and blink, then fall asleep in whichever lap they happen to have landed on.

Despite the excitement of the babies, Chloe seems subdued today. As the younger kids crouch over the tiny goats, Chloe hangs back, her tall, thin frame stooped as she watches the action from above. When I ask Andy whether something is bothering Chloe, she explains that Chloe has had a rough month. At Chloe's magnet high school, the college admission process consumes kids in their junior and senior years. Some families hire private counselors to help kids navigate the admissions process. Many get assistance with their essays, but Chloe has stubbornly refused any help. Back in the fall, she told me that she thought working with a private counselor would be unfair to students whose families couldn't afford one.

Chloe sent in her applications and waited. When the letters started arriving last month, the news was not good: Chloe was waitlisted or rejected at almost every school she applied to. She was admitted to one of her safety schools—a small college in Washington State called Whitman—and several University of California schools, including Berkeley. She knows Berkeley is an excellent school, but she grew up in its shadow and was hoping to try living somewhere else.

When I have a moment alone with Chloe later in the day, I ask her how she is feeling. She says with a sigh that she is sure it will all turn out fine. She recognizes how lucky she is to have a keen academic mind that her rigorous high school has sharpened—and a supportive and encouraging family that can afford to pay for college. Still, the sting of the rejections was much sharper than she expected.

Showmanship practice is a good distraction. Chloe manages to tear the group members away from the babies and assigns each one a bigger goat to practice showmanship with. The first task is to "set up" the goats, spreading their back feet apart slightly and lining them up with the front feet. Chloe explains that the goat's head is held at a ninety-degree angle from the body, and its back must be flat. If the goat begins to arch its back, you put a little pressure on it with your hand so that the muscles flatten out. Everyone works on setting up, and Chloe makes the rounds to each child and goat, giving pointers as she goes.

"Always make sure the goat is where?" she asks.

"Between you and the judge," chorus a few of the kids.

"How do you remember the parts of a goat if the judge asks you to name them?"

Silence. Everyone seems absorbed with getting their goats to cooperate. A few goats keep making a break for the pen. There is a lot of running and squealing.

"Winning champions love root beer!" calls Chloe, above the din. "And does everyone remember what that stands for? Withers, chine, loin, rump."

Chloe has everyone walk in a circle, as they'll do in the show ring. She offers the group a few tips: don't pull too hard on the leash, and when you do pull, pull up, not back, so you don't end up strangling the goat.

The next task is to take the goats on a short walk, so they'll get used to being on the leash. We head down the driveway, crossing the street by the fire station. The girls manage the goats with varying levels of success. For some reason, the smallest girl in the group has been paired with Nutmeg, the biggest doe. Most of the time, Nutmeg seems to be the one running that show.

On the way back, Asher, the only male goat that has come along on the walk, starts misbehaving, bucking and twisting out of his collar. He probably weighs a hundred pounds, more than most of the kids. I know that Chloe could handle Asher, but the group has fanned out a little, with Chloe in the lead way up front. I'm in back, and I can see that the girl who has Asher is getting rope burn from trying to control him. He bucks up on his hind legs and pulls at the leash with all his might, and the girl pitches forward. This street has no sidewalk, just a raised path. There aren't many stoplights up here in the hills, and cars go fast. I decide to intervene, taking the leash.

Asher calms down for a minute. We walk a few paces. Then, he strains with all his might and breaks into a bolt. I can't believe how strong he is. I'm thrown off balance, and all of a sudden, I'm being dragged on the ground as the little girls gape. Chloe must have seen the commotion, because instantly she's there, taking Asher's leash from me and apologizing.

Figure 12. Allison and Sydney play with their pygmy goats at Borges Ranch in Walnut Creek, California. Photo by Rafael Roy.

"I am so sorry, Kiera, he is such a butt. Are you okay?"

I assure Chloe that I'm fine and smile at the girls around me, hoping they don't notice my cheeks turning bright red. Somehow, Chloe has had a calming effect on Asher, who is walking quietly beside her. When he does jump or squirm, Chloe throws her hip against him and talks to him reassuringly. She walks him all the way back to the pen, and I trail sheepishly behind, dusting the dirt off my jeans and wondering whether I'll get a bruise on my throbbing knee.

On a gorgeous day in late May, I drive out to Borges Ranch for the Pleasant Hill goat group's last meeting before the fair. Spring is in full force at the ranch. Wildflowers dot the hillsides, and the trees and the grass are lush. On my way up the path from the parking lot I see a baby rattlesnake slither off into the brush.

In a paddock by the barn, Allison's best friend, Sydney, is sitting on the ground with her goat, Axel. Alyssa and Kayla, the younger members of the goat group, are taking turns holding Butterball's leash. Axel and Butterball have grown a lot; they are at least 25 percent bigger than they were the last time I saw them.

KC comes out of the barn and says that Allison is down by the pen, retrieving her goat, Rayne. KC tells me that it's been a good few weeks. Allison has been working on showmanship with the other girls, and she still loves her independent study program at school, especially her art class. She's been sewing a cloth doll in a miniature version of the camo prom dresses that KC has made for her and Sydney. KC loves watching Allison get lost in art projects for hours at a time. "She can focus if she wants to," says KC.

Allison takes Rayne into the paddock to join the other girls and their goats. Allison's nails are even more spectacular today than usual, with gold tips. She's also wearing artfully ripped jeans.

"I'm going to pretend to be the judge," says Allison, moving to the center of the paddock. "Okay, walk," she commands. She turns slowly in a circle, giving pointers, encouragement, and corrections as Sydney and Kayla walk their goats around her.

Allison walks over to Kayla, stoops down, and puts her hands on Butterball's back end and tousles his fur. Then she stands and looks at Kayla expectantly.

"What?" says Kayla.

"What are you supposed to do now?" asks Allison. "Do you remember?" Kayla shrugs.

"When the judge messes up your goat's fur, you're supposed to smooth it back immediately," she says. "Pretend the judge has a disease, and you have to wipe it off your goat."

"Oh yeah," says Kayla. She pats Butterball's backside, and Butterball bucks.

"Don't hold his leash like that," chides Allison. "Gather it up so it's shorter. Now grab his butt." Allison takes the leash, arranges Butterball, and then hands the leash back to Kayla. Butterball stands still for a minute, and Kayla tries to position his legs. He jumps and backs up. Kayla sighs in frustration and looks at Allison.

"If he does that, don't freak out," Allison tells her. "Don't let him take over; otherwise, he's going to think he can get away with everything. You have to show him that you're the boss."

Sydney takes a turn being the judge so that Allison can practice with one of the other goats, Chewy, who seems to want to lie down more than anything else. Allison isn't worried. "He's just being like this because he wants to go back to the barn," she says.

Alyssa practices with Rayne, looking slightly bored. It's hard to tell if she's really interested in the goats, or just in hanging out with the big girls, Allison and Sydney.

Allison comes over to the picnic table where I'm sitting with KC. "You were great out there helping Kayla and Sydney today," I say. I ask her if she's nervous.

"I wouldn't say nervous," Allison replies. She says she wants to go to a showmanship clinic in a few weeks so she can work with other species, just in case she makes it into the master showmanship competition. "I just really want to win small-animal masters this year," she says. "So I have to study a little before I go."

Kelly and her brother, Kyle, are also deep in fair preparations. I arrive at the Semonsen sheep farm one morning to find them setting up two red metal contraptions. Each has a grated platform and, on one end, a post that stands about two feet high at right angles to the platform with a short chain dangling from the top. It looks like a medieval torture device. Kelly looks up from securing the post and waves to me quickly, then returns to her task.

"What are those things?" I whisper to their father, Steve, who's fixing something by the barn.

"They're stands," he says. "This is a trick that they learned at lamb camp." The lamb, he explains, stands on the platform, its chin goes on the bar, and the chain goes behind its head. It's good for lambs who haven't been trained yet, because it keeps them from running away. It looks scary, but it actually keeps the lambs calm and makes them feel secure."

"Lamb camp?" I ask.

Steve explains that this is a show clinic held every spring a little more than an hour away in Dixon, California. Kyle won reserve grand champion showing one of the sheep that Kelly had trained, while Kelly went home empty-handed.

"It was really frustrating for Kelly, because she had done all the work with the lamb, and her brother got to get up there, had his picture with the judge, took home that trophy and that embroidered jacket. It was a very quiet ride home." Kyle, Steve says, cares much more about soccer than about sheep, but he's a natural at showing lambs. Instead of becoming discouraged, Kelly has worked harder than ever with her lambs this week. Steve says that's typical of Kelly: setbacks only strengthen her resolve to win.

Kelly disappears into the barn for a few minutes. When she emerges, she's hustling a large lamb out of the barn and onto the stand. The lamb, named Leonard, is obviously not happy about going on the stand. He struggles, and Kelly shifts her weight, thrusting out her hip to block him from bolting. Once she has him next to the stand, she hoists him onto the device and secures him with the chain. Almost immediately he calms down.

Today, Kelly is wearing jeans and an old T-shirt, an outfit that allows for easy lamb wrangling. Her blonde hair is pulled back from her face in a low

Figure 13. Kyle Semonsen, 14, practices with his lamb at his home in Castro Valley, California. Photo by Rafael Roy.

ponytail. Unlike Allison and Sydney, Kelly does not bother with stylish clothes and makeup when she's working with her animals.

Next, Kyle leads his lamb out and gets her up on the stand, though with a little more struggle. Kyle is small for his fourteen years—I'd guess eighty pounds dripping wet—so he has a harder time than Kelly controlling a hundred-pound sheep.

As Kelly and Kyle jostle around the stands, setting up the lambs' feet so that they're pointed straight ahead and just the right distance apart, I ask Kelly what the judge looks for in a lamb. "You want your lamb to look classy," she says.

"Classy?" I imagine a lamb wearing a string of pearls. "Please explain."

"Well," she says slowly, "you want one that's very triangular in shape, meaning the shoulders and chest up front are smaller than in the back where the hips are. You want one that's very level across the top so it's smooth all the way. You want one that has a longer neck, one that's not as sharp in the shoulders, so there aren't those little points. You want one that's very wide across its ribs. From the thirteenth rib—that's the last one—to the hipbone should be long, as long as possible. You want its feet to be facing front instead of out to the sides or in. Am I missing anything?" Kelly turns to Steve, who has been listening quietly from the other side of the fence.

"You could tell her why you want a long neck," he says. "It's a growth indicator. Means the lamb will grow big."

Kelly splays Leonard's back feet apart on the platform. He picks a hoof up and shakes it, and Kelly puts it back where it was.

"I used to do this part just holding them, but at lamb camp I learned that it's helpful to use the stand so you don't have to worry about them running away and you can focus on getting them used to you touching their feet."

Once the lambs are comfortable being set up, Kelly will practice bracing them. The brace, she says, is one of the most important lamb maneuvers. Facing the lamb, you throw one leg forward and put the lamb's chin on your upper thigh. The idea is to get the lamb to lean into you so that its back muscles are taut. "You want them to do that so that they flex their muscles and the judge can see them," says Kelly.

After a few more minutes of working with Leonard on the stand, Kelly unchains him and takes him back into the barn. Next she brings out a smaller sheep named Penny, who has become her pet. Penny often tries to get Kelly's attention by chewing on her pants. In some lambs, says Kelly, that behavior would become bothersome, but for some reason in Penny she finds it endearing. Kelly has decided not to enter Penny in the auction, so she will come home with Kelly after the fair.

As Kelly is finishing up with Penny, I ask if she thinks she will feel sad when it comes time to say goodbye to Leonard and Walter, the two sheep she will take to auction at the fair. She shrugs. "Maybe Walter, a little bit. But it's

Figure 14. Kelly braces a lamb in the practice ring on her family's farm in Castro Valley, California. Photo by Rafael Roy.

my mom who always cries." In general, Kelly is not sentimental about the animals. She feels proud of how she and her family treat their lambs, giving them plenty of space for exercise and fresh grass to graze. And anyway, by the time fair comes around, she's usually fed up with the responsibility of working the animals for upward of an hour and a half every day, so that helps make parting easier. The promise of auction money doesn't hurt, either.

Watching showmanship practice helps me understand some of the show conventions. I now get why it is so important to set up your goat or sheep—if it was just running around, the judge wouldn't be able to evaluate its muscles. But some of the other showmanship conventions still seem sort of silly to me. Why does it matter which side of your animal you stand on, or whether your goat's head is at a ninety-degree angle?

These things begin to make sense a few weeks later, when I attend a class called Quality Assurance and Ethics Awareness at the Alameda County Fairgrounds, just before fair season. All 4-H'ers who are showing animals for the first time are required to take it, so I tag along with some of the girls from the Montclair goat project. The teacher moves swiftly through rules about what exhibitors are allowed to give their animals before the fair, most of which are obvious: you can't force-feed your animal, for example, nor can you give it off-label medication without a prescription from a vet.

But the teacher also talks about showmanship. The rules were designed to help exhibitors show "to the animal's advantage"—so that potential buyers could assess the amount and quality of meat. At the fair, showmanship serves another purpose, as well: it helps win the trust of the public. The teacher reminds the children in attendance that they are representing agriculture for the half million visitors who come to the fair annually. It's possible that the fair will be the visitors' only agricultural experience all year long. Do the kids want the visitors to come away with the impression that farmers don't care about the health and well-being of their animals?

I think back to one of my earlier visits with Anthony. We were standing just outside the barn with the animals when, from the far corner of the pasture, we heard someone yell, "Hey! Can I meet your pigs?" A woman in jogging gear was standing by the fence, gesticulating. Anthony rolled his eyes. "It is so annoying when this happens," he muttered. Nevertheless, he used his pig stick to drive Panda and Domo over to the woman, who let loose a steady stream of coos. "Oh my god, they are sooooo precious!" said the jogger. "Hi babies! They look like babies, how old are they, they are *just the most precious* things I've ever seen, aren't they?"

Anthony nodded and said something like "Yeah, sure." I felt indignant on his behalf. He had real work to do with the pigs, not to mention homework waiting for him. Did this woman think his pigs were just cute animals there for her viewing pleasure? But when we got up to the barn, Anthony's mom, Susan, said, "It's always good to be polite, since we are raising these animals in a neighborhood, and we don't want to get a bad reputation."

Showmanship is kind of the same idea. You gussy up your animal and follow all the rules at the show so that the fair-going public can see how serious you are about taking care of this creature. The show audience hasn't been in the barn with you every day as you measure your pig's feed or halter train your lamb, so you have to make all that hard work visible in your every gesture during the show.

CHAPTER EIGHT

Grow 'Em Big

Five days before the fair, it's time for the kids in Anthony's pig project group to weigh their pigs. They're supposed to keep track of their animals' weight throughout the project, but not all have regular access to a scale. Every few weeks, Sally (the pig group's leader), her father, and another helper have put a livestock scale in a pickup truck and driven it around to each place where the members of the group keep their animals.

Pigs of any size are allowed to compete at the fair, but in order to qualify for auction they must weigh between 215 and 280 pounds. In years past, the fair allowed 4-H'ers to sell their nonqualifying animals privately and to transport those animals on the same slaughter truck with the pigs that were auctioned. But this year, the rules have changed: only auctioned animals get a free ride to the slaughterhouse.

That means the stakes for making weight are high: kids whose pigs are too light or too heavy will have no chance to earn back the money they spent raising the pigs—usually several hundred dollars per animal. (There are weight requirements for all market species, but the pressure to make the cutoff is greatest for pigs and steers; they are heavier than lambs and goats, so families spend more on their feed.)

I drive out to the suburb of Martinez to see the pig group's weigh-in. It's at a central barn where many of the kids in the group keep their pigs— the barn that the kids got into trouble for not cleaning at the last meeting.

The sun is setting, and the hillsides on either side of the highway are turning golden. When I arrive at the barn, I see Sally standing in the yard, clipboard in hand. I remark that the place looks tidy today. Sally raises her eyebrows and pulls me away from the cluster of parents standing around outside the fence.

She fills me in. "Things got bad again last week. I had to send an angry e-mail, which means it was really bad, because I don't send angry e-mails." As a result, one mom isn't speaking to her. This, says Sally, is only adding to the tension on this already tense day.

Sally thinks most—but not all—of the pigs are on track to make weight. "We have a few lighties that I'm worried about," she says.

Kids are chasing their pigs into the barn, while their parents look on from behind the fence. In the barnyard, some older men—including Sally's father and uncle—are setting up the scale, which looks roughly like a pig-sized cage, with its open side positioned toward the barn door. It takes several kids to wrangle a pig out the barn door and onto the scale.

I chat with a friendly woman named Jamie, the mother of Jackie, a tall twelve-year-old with long, dark hair and glasses. We watch Jackie chatting animatedly with some of the other girls as they wait for their pigs' turn on the scale. At last weigh-in, says Jamie, Jackie's pig wasn't on track to make the cutoff. She tells me that Jackie has been feeding the pig nine or ten cupcakes a day for the past two weeks, a tried-and-true calorie-delivery method for underweight pigs. Jackie has also been adding a fat supplement called Sumo to the batter to cram in more calories.

When it's Jackie's pig's turn to be weighed, Jackie shoos him out of the barn and onto the scale. One of the men announces that he's 204 pounds.

"He has to gain two pounds a day for the next five days," says Jackie's mom, shaking her head. "That's not going to happen." She picks a bag of cupcake wrappers off the ground. "What am I going to do with this pig?" Jackie is still clowning around with some of the other kids, showing no outward signs of concern or disappointment. "Hey, Jackie," calls Jamie. "He's 204."

"Yeah, I know," says Jackie. Then she turns back to her friends. I remark that Jackie seems to be handling this well.

"She'll cry in the car," says her mom. "I know that face."

The next few pigs step onto the scale. All of them make weight except for one, a 198-pound female that belongs to a younger girl with bright red hair. Sally talks to the girl and her dad afterward. "She'd have to gain more than three pounds a day," says Sally. The girl lowers her gaze. She doesn't say anything, but stands back with her dad as the other kids bound out of the barn.

Once all the pigs at this barn have been weighed, the men load the scale onto the pickup truck. The next stop is the barn where Anthony keeps his pigs. Even though Anthony has his own scale, Sally wants to double-check his pigs' weight with her scale.

Anthony and his mother and grandfather are waiting, and the pigs— Panda and Domo—are in the pen with Satchel the lamb. Sally's father and uncle lift the scale out of the truck and put it in the barn. Panda, who is younger than Domo, is in the 180s, so she won't go to auction at this fair (though she'll still qualify for showmanship). That's okay with Anthony; he had planned all along to enter her in another fair later in the summer. Domo weighs in at around 250. When he steps off the scale, I watch him shuffle around the pen. I can hardly believe this hulking creature is the same Domo that I met just a few months before, when he was barely out of piglet-hood.

When I first began hanging around with 4-H'ers, I noticed that everyone was obsessed with what they fed their animals. Kids and parents debated not only the merits of various feed brands but also the picayune details of how and when to present the food to their livestock. Before or after exercise? Mixed with water or left dry? Their fixation on food rivaled that of diners in trendy California restaurants.

I didn't understand what the fuss was all about. Couldn't you just go to the store, pick up a bag of feed, and read the directions? I posed this question to Kelly one day.

"You have to keep very careful track of what they're eating at the different phases of their lives," she explained. All lambs start out nursing. Kelly gradually introduces solid food after a few weeks, a practice known as "creep" feeding. When the lambs are very small, she sets up the feeders in such a way that they can get in to eat anytime they want. Only once the babies are fully weaned does Kelly begin weighing them every day. She starts them on a daily diet of 3 percent of their body weight in feed, divided between two feedings per day. That way, ideally, they gain about half a pound every day, and Kelly adjusts the rations according to how much they're gaining.

Some people feed lambs different mixes of food at different points in their lives, but Kelly uses the same food throughout: around 18 percent protein, with oats, barley, and corn for texture and flavor. "That way it's more like granola than lima beans," she said. "So they don't get tired of it."

Then there's the presentation of the food. Kelly explained that she likes to add water to the dry feed. The lambs seem to like it better, and it makes it easier to add in extra nutrients, such as electrolytes, so the lambs stay well hydrated. Kelly doesn't believe in feeding her lambs too many supplements—her method of careful feed monitoring and exercise has always produced excellent results. But some people, she said, feed all kinds of commercial additives whose producers say they will help animals put on fat and muscle.

Few 4-H'ers use these products for sheep, but when it comes to pigs, growth-enhancing supplements are increasingly common. Sally attributed this trend, at least in part, to a change in judging standards: over the past few years, judges have begun to favor larger, more muscular hogs. "You see it more and more in 4-H," Sally lamented. "The idea that you have to grow this Arnold Schwarzenegger pig in order to win, and the only way you can get it to be big enough is to feed it chemicals."

Growth-promoting supplements have evocative names. Offerings from a feed producer called Sunglo include Explode, Inhale, and Full Tank.[1] On the website of a brand called Essential Show Feeds, animal owners can choose from a dizzying array of supplements, each one promising a different effect.[2] For pigs, Wide Open ($49 for 42 pounds) "expands the ribs & opens the body—Does this while maintaining normal weight gain." A similar product for lambs, X-Pand-O ($20 for 25 pounds) "produces an expanded look and keeps your animal fresh." Keep'n On ($5 a bottle for goats) "provides muscle pop, energy for the arena, fill without the weight and 12 vitamins and 9 minerals to help stamina and recovery."

Of the many supplements that 4-H members feed their livestock, a chemical compound called ractopamine hydrochloride, which promotes muscle growth when fed in the final weeks of an animal's life, is probably the most controversial. Ractopamine is part of a group of chemical compounds known as beta-agonists that mimic the hormones epinephrine and norepinephrine, thought to trigger the fight-or-flight response in mammals. Developed in the 1990s, when the low-fat diet fad increased demand for lean pork, ractopamine is currently approved under the name Paylean for pigs, Optaflexx for cattle, and Topmax for turkeys. Among US pork producers, it's an extremely popular drug; an estimated 60 to 80 percent of the pigs in the United States are treated with Paylean.[3]

Ractopamine is manufactured by Elanco, a company owned by the pharmaceutical firm Eli Lilly and Company. (According to its charitable giving brochure, Elanco also supports 4-H.[4]) Kathy Vannatta, who has worked on research and development at Elanco, explained that ractopamine helps an animal "redirect its nutrients to muscle rather than fat" so that the animal grows bigger in a shorter period of time. "It has been a key part of farmers making the most of the feed that they use," said Vannatta. "Grain is very expensive right now, so feed is a big cost. A quarter of American families struggle to put food on the table, so we believe that ractopamine is a valuable tool in fighting hunger."

When the FDA approved ractopamine for pigs in 1999, the agency relied on safety tests conducted by Elanco. But shortly after the drug went on the market, the FDA began hearing reports of pigs reacting badly to it. In 2002, the agency sent a letter to Elanco reprimanding the company for failing to submit several key pieces of information, including customer reports of adverse effects and data from the company's own initial safety tests.[5] Later that year, the FDA began requiring Elanco to attach a warning to its labels: "Ractopamine may increase the number of injured and/or fatigued pigs during marketing. Not for use in breeding swine."

Jeremy Marchant-Forde, a research animal scientist with the USDA's livestock behavior lab at Purdue University and an expert on pig welfare, has studied the effect of ractopamine on pigs. In 2001, the supplement had been on the market for just two years, but its sales were skyrocketing. Marchant-Forde and a team of graduate students had been "getting reports from farmers that pigs on ractopamine were harder to handle," he recalled. His group launched a study to see whether the compound affected pigs' behavior, tracking seventy-two young pigs, half of which were given ractopamine in addition to their regular feed during the month before they were slaughtered.[6]

From the first week, Marchant-Forde said, the ractopamine-fed pigs were "more alert and spent more time at the feeder." They were observed to run around the pen more and lie down less than the control group. By the second week, they began to resist leaving their home pen to go to the weighing scales—which the control group usually did voluntarily. They required more "pats, slaps, and pushes" from handlers when being moved. During the week before the pigs were slaughtered, Marchant-Forde also found that the ractopamine-fed animals had higher heart rates and levels of epinephrine and norepinephrine than the control group.

Since that first study, Marchant-Forde and his group have continued to investigate ractopamine's effects on pigs. In 2008, they found that ractopamine-fed pigs developed more hoof lesions than a control group. In 2009 and 2010, they documented more aggressive behavior and higher levels of stress hormones. The same group had lower levels of the neurotransmitter serotonin, which stimulates feelings of calm and contentedness.[7]

The FDA's Center for Veterinary Medicine documents cases of livestock injury, illness, and death that could be related to drugs. An investigation by Helena Bottemiller of the Food and Environment Reporting Network found that between 1987 and 2011, the agency documented 218,116 cases of ractopamine-fed pigs experiencing adverse effects, including "death, recumbency, lameness, hyperactivity, reluctance to move, stiffness, trembling, dyspnea, collapse, and hoof disorder." More than six times as many adverse effects were reported for ractopamine than for any other drug.[8]

While the FDA said that it doesn't have enough information to determine whether the drug caused any of the adverse effects, pork producers' accounts are striking.[9] One from an Indiana farmer in April 2011 noted that "pigs in a research barn squeal when they take steps, as if in pain. . . . This is an ongoing issue at the site w/ractopamine-fed pigs. Handling has not changed and no new procedures had been introduced."

Many note an increased incidence of "downer" pigs—those that are unable to walk—among those fed ractopamine. This 2010 incident report about an Iowa producer is typical: "The producer reported that the pigs that were finished on Paylean would be 'so stressed' while being moved or sorted that the pigs would lie down and die. . . . The producer fed Paylean to his finishing pigs intermittently from 2005–2007 and reported the same experiences each time he fed Paylean. The producer estimated he lost approximately 40 pigs over the 2 year period."

In addition to raising questions about ractopamine's health effects on animals, some producers have voiced concerns about whether traces of the drug remaining in the animal at the time of slaughter might affect people who eat the meat. While there are no conclusive studies about ractopamine's human health effects, several US food companies, including Chipotle, Whole Foods, and Niman Ranch, refuse to buy animals that have been treated with the drug. Ractopamine has been banned in China and the European Union, and China's refusal to accept US pork treated with Paylean kicked off a trade dispute, Bottemiller reported.[10]

In general, Jodi Sterle, an associate professor and swine nutrition expert in Iowa State University's Department of Animal Science, doesn't think Paylean is detrimental to the health of pigs if used judiciously. Sterle, who has helped develop food safety curricula for junior livestock exhibitors, pointed out that both 4-H leaders and county fairs go to great lengths to make sure that kids understand that not following label instructions on supplements is illegal. But Tom Baas, an Iowa State 4-H swine specialist who has judged hundreds of youth pig shows over the past two decades, estimated that 90 percent of 4-H'ers who raise pigs use Paylean. He said he suspects that because of the competitive nature of the shows, not all of them follow the label's instructions. "If someone believes that they are going to gain an advantage by feeding more Paylean, I'm sure it's going to happen," he said. "I've seen pigs that tend to be much more irritable and have movement problems, and my suspicion is that they may have been fed more Paylean than the label allows."

Sarah Watkins, a 4-H program representative for animal science at the University of California's Agriculture and Natural Resources division, guesses that only a handful of California 4-H'ers feed more than the recommended dose of ractopamine; she pointed out that not following label instructions is grounds for disqualification both in 4-H and at most fairs. However, ractopamine doesn't stay in an animal's system for very long, and there is no federally required withdrawal time for it before slaughter—so it would be almost impossible for a fair to determine whether an exhibitor had overused it. "There is pretty much no way to trace it," said Watkins.

In Martinez, Sally discourages the members of her pig project group from using Paylean. At the restaurant where she works, an increasing number of customers ask whether the meat that the restaurant sells comes from animals raised with chemicals or hormones. She worries that the ractopamine culture in 4-H is out of touch with dining trends—especially the new guard of consumers who prefer naturally raised meat. Anthony says he doesn't feed ractopamine; he prefers the time-honored approach of supplementing show feed with cake mix for pigs that don't gain.

Even for 4-H'ers who don't use ractopamine, it's almost impossible to raise an animal without any drug additives. Most commercial feeds today contain antibiotics, chemical compounds, and mysterious "proprietary blends." The show feeds that 4-H'ers use are no exception.

Honor Show Chow is the feed brand preferred by many of the 4-H'ers I met. I visited the company website, where a link to a section called "Feeding Strategies" brought me to a list of different species: pigs, lambs, chickens, goats, and steer. I clicked on the pigs section, which allowed me to choose the level of muscle I wished to achieve in my pig. I settled on "heavily muscled barrow" (castrated male), which recommended a regimen of five different feeds over the course of the pig's lifetime.[11]

I clicked around on the site. All the feeds had protein, fats, and various vitamins and minerals—and some extra ingredients that I didn't recognize. Showpig 509 included a proprietary blend that promised "full middles, muscle shape, and bloom." Almost all the feeds contained an algae-based additive called Tasco®, which, according to the site, helped "lower body temperatures during periods of heat stress helping to keep pigs on feed during warm weather." Showpig 509 and Showpig 709 were available with a dewormer called Safe-Guard® and the antibiotics lyncomycin and tylosin to prevent various diseases.

In the lamb section of the Honor Show Chow site, I found more unfamiliar ingredients. Most of the feeds contained a medication called Deccox® to prevent the parasitic disease coccidiosis, a yeast culture called Diamond V©, and a mineral mix called Zinpro Performance Minerals®.

The livestock industry has drawn a growing amount of criticism for its heavy use of additives, especially antibiotics. An incredible four-fifths of the antibiotics sold in the United States today goes to livestock, some of it in the form of feed additives.[12] That's because low doses of antibiotics have been found to promote growth in animals.[13] Some public health experts worry that the use of antibiotics on farms could contribute to antibiotic resistance in humans.

University of California's Watkins doesn't see a problem with medicated feeds. "Those low doses of antibiotics help promote growth because they prevent infection, and if the animal isn't having to fight off infection, it has more energy to put toward growing," she says. In market animals, "from the 4-H side, I have never heard of people having concerns about low doses of antibiotics." Still, many of the 4-H families that I met said they were concerned about additives. Some said that they had considered switching to organic feed, but it was simply too expensive.

One family I came across, however, found that their initial investment in additive-free feed paid off. In 2012, seventeen-year-old Valerie Pors and her fifteen-year-old brother Abel, 4-H members in Charlottesville, Virginia, decided to raise their two lambs on organic feed, grass, and hay.

"There are a lot of people out here who are bothered by how 4-H pushes toward the weight gain whether or not it's good for the animal," recalls Joyce Pors, Valerie and Abel's mother. "Some of that bothered us." The family's vet, she says, "didn't agree with the 4-H recommendations for feeding, because of the digestive stress on the animals. They recommend feeding lots of grains for weight gain, when their natural diet is grass and weeds."

The Pors's lambs gained enough weight on the organic feed to be shown at the fair, but Joyce didn't have high hopes for the show ring. "We didn't expect them to place," she says. She also didn't expect to recoup the costs of raising the sheep, since organic feed was significantly more expensive than conventional.

And they didn't place. Out of a market class of six lambs, they came in fifth and sixth. That was okay with the Pors, because their "goal was not to win; it was to learn how to raise lambs," says Joyce. She didn't expect them to earn much at auction, either, since the price that livestock fetches usually corresponds to how well it does in the ring.

But on auction day, something surprising happened. When Valerie took her fifth-place lamb onto the auction block, bidding started off low. The auctioneer was about to call it when an acquaintance of the Pors's made her way to the podium and whispered something to the auctioneer. Then, the auctioneer turned to the crowd and announced that Valerie's lamb had been raised on organic feed.

Bidding took off, and Valerie's lamb ended up earning the highest price so far that day. Abel's sixth-place lamb went next and did even better—in the end, it fetched more than the grand champion did. The family even ended up making a modest profit on both animals—despite the relatively high cost of organic feed.

The Pors are not alone. The Thode family in Sebastopol, California—the people who sold me the two turkeys I raised—has also figured out a way to make organic 4-H projects profitable. I didn't know it when I bought my birds from them, but the five Thode kids raise dozens of turkeys every year as 4-H projects. Soft-spoken twenty-one-year-old Zach is the veteran turkey farmer of the family. In 2004, he began raising heritage turkeys—which resemble wild turkeys and have a gamier flavor than most turkeys sold at supermarkets—for their 4-H club. Through the club, the kids tried to sell their birds for Thanksgiving, but they had a marketing problem: their customers wondered why they should pay $6.50 a pound when supermarket turkeys cost a fraction of that.

In 2007, the local Slow Food chapter got wind of the project. Jim Reichardt, a veteran poultry farmer and member of Slow Food Russian River, had been trying to figure out how to save heritage turkeys, as few people raised them anymore. "We were so impressed with how these kids were raising their birds," he recalls. "We wanted to support the project." It was a natural match: the 4-H club had the birds, and Slow Food had the network of foodies to buy them. Slow Food members even negotiated with a local feed mill to make a custom organic feed for the 4-H'ers' birds. The project has grown every year since. In 2012, eleven kids raised 225 birds. Project leader Catherine Thode estimates that some kids earn upwards of $1,000. A few years ago, Zach spent his earnings on a used pickup truck.

So why aren't organic 4-H projects taking off in more places? The reason likely has to do with local culinary trends. The Pors's and the Thodes' projects work because where they live, people are willing to pay more for organic meat. Charlottesville, Virginia, and Sebastopol, California, have well-established culinary communities. It's unlikely that similar projects would be as successful in places without nearby Whole Foods and regular farmers' markets, says Bill Ahrens, owner of Countryside Organics, the store where the Pors bought their organic lamb feed. "In general, I don't know that there is a great penetration of organics into 4-H," Ahrens wrote to me in an e-mail. University of California's Watkins agrees. She doesn't think that organic 4-H livestock projects would be economically feasible in most parts of the country. "Right now, buyers that attend county fairs are more into supporting the youth, and not necessarily eating that animal," she says. "The whole farmers' market spiel—I don't think that would really work right now, at least not in all places. Organic feed would be super-expensive, so you would need to have a mark-up price."

But the Pors's experience suggests that, at least at one fair, there is a growing interest in meat raised without chemicals or hormones. "It was really interesting, because a lot of these guys who were bidding on our lambs are more traditional farmers," says Joyce. "I guess they must have liked something about what we were doing."

Sally says that she looked into organic pig feed a few years ago, but the only product she found was made in Canada, and the cost of the feed and shipping would have been exorbitant. Raising a pig just on organic grains and vegetables was an option, but she knew that without a reliable protein source, it wouldn't gain enough muscle mass to compete with the other show pigs. (Because they are not grazers, pigs might be harder to raise organically than ruminants like cattle, sheep, and goats, which can obtain much of the nutrition they require from organic grass.)

Even if organic fair pigs are unrealistic, Sally would like to see the fairs offer "commercial" pig classes, for animals raised on standard feed rather than the souped-up, expensive show stuff. "This way kids could raise pigs on commercial grade food and have a good, well-fed pig that may be 'ugly,'" she explained to me in an e-mail. But she suspects that even that would be difficult, since at auction, these commercial pigs would still be competing against the much more muscled show pigs. What's more, without show feed, the commercial pigs might not gain enough weight even to qualify for auction.

"Unfortunately, the pig show at the fair really is a pretty pig contest these days," wrote Sally. "No one is willing to admit that the fundamental purpose in teaching kids to raise animals, and to produce animals for market and a sustainable community, got lost somewhere."

But there's another reason that we might want to make sure that 4-H'ers feed their animals only food that we feel comfortable eating. At most fair auctions, buyers can choose to "resell" the animal back to the slaughterhouse instead of taking the meat home. This option is popular among people who want to support 4-H or Future Farmers of America but don't have the freezer space or desire to keep, say, an entire steer at home. The slaughter-house then sells that 4-H meat to its regular customers, which may include major grocery chains and restaurants. This, of course, means that the chops that you pick up at the supermarket for dinner could have started off as a champion 4-H hog.

CHAPTER NINE

The Contra Costa County Fair

Fair season has finally arrived. June 1 is the opening day of the Contra Costa County Fair, where Allison will show her pygmy goats and Anthony will show his pigs. Three weeks later, Chloe, Serena, and Kelly will head to the Alameda County Fair.

Because of different fairs and different species, none of my 4-H'ers will be competing against one another. But how will they—and the rest of their groups—fare against the other competitors? By this point, I know the crew well enough to make a few predictions. At the Contra Costa County Fair, I guess that Allison will take first or second prize in showmanship—she's been studying and practicing constantly up at Borges Ranch, and her mom, KC, has been quizzing her on curveball goat questions. I'm willing to bet that Anthony's pig Domo will place in the market class—to my untrained eye, at least, that pig looked large and in charge at weigh-in the other night. I'm curious to see how Anthony does in showmanship. I know he's taken ribbons home in the past, but he's been so shy with me that I can imagine him having a tough time enduring rounds of questioning from the judges.

At the Alameda County Fair, I predict that Chloe and Serena will do well in dairy goat showmanship, and I am confident that their Montclair 4-H Rabbit Bowl team will remain undefeated. Their goats, however, likely will be no match for their more fussily bred opponents.

Figure 15. Anthony waits for a show to start at the Contra Costa County Fair in Antioch, California. Photo by Rafael Roy.

And Kelly? My imaginary money is on her as the big winner. I wouldn't be surprised if one of her sheep takes a first place, or even a grand champion, and I expect that all her focus and dedication with the lamb stands on her family's sheep farm will pay off when she takes first in showmanship. Maybe she will even win master showmanship.

Tonight, the eve of the Contra Costa County Fair, is livestock weigh-in. Any animals that don't make weight on the official fair scale won't be allowed to compete in the market class. Weigh-in, I'm told, takes approximately forever. Not just pigs but all the market species—lambs, steers, meat goats—have to be weighed. When I asked Anthony's pig group leader, Sally, if I could come see the weigh-in, she said, "You want to come sit on fences with us? That's what it's like: kids sitting on fences and waiting."

"I'm in," I said.

The Contra Costa County fairgrounds are in Antioch, California, a small city about forty-five miles northeast of San Francisco. The housing bust of 2008 hit this city hard, and it hasn't yet recovered; housing values are half what they were a decade ago, and in some parts of the city, foreclosure rates are more than double the national average.[1] I pull off the highway into a car-dealership district. At the end of the busy road, there are empty storefronts and a sign for some fast-food places. This, I think, doesn't exactly look like the idyllic home of a county fair.

But once I turn into the fairgrounds, it's a different world. Workers are constructing what will become the midway; trucks slowly transform into spinning teacups and Ferris wheels, like the Transformer toys I remember from childhood.

The livestock barn is way in the back. An old wooden building, it's lined with pens freshly decorated for this year's fair theme, Here Comes the Sun. Parents and 4-H leaders have told me that their kids seem to forget the world outside when they're in the fair barn, and now that I'm here, I can see that it has a timeless quality. Recently I came across a series of interviews conducted for an oral history of the Extension system. One of the people interviewed was Raymond Lyon, a seventy-five-year-old former 4-H leader from California's Central Valley.[2] The interviewer asked him how fairs had changed over the years, and he responded: "My wife and I judge at a lot of county fairs, so we go around, and the kids look exactly—not just their dress but their actions—they're exactly what they looked like thirty and forty years ago. . . . It's really unusual, how it's the exact same things. You could come back after—and you'd fit right in, you know."

As I enter the barn, I see Anthony's mom, Susan, hustling out toward the parking lot. I wave hello, but she is clearly flustered.

"Anthony's pig is bleeding," she says. "I'll catch up with you inside."

Right away I see Anthony bent over Domo, who, as Susan warned me, is bleeding—a lot.

"They did his ear tag wrong and had to redo it, and they hit a vein," says Sally, who is there with her clipboard.

I glance at Anthony, who looks as calm and unfazed as ever, holding paper towels up to Domo's ear. Susan comes back.

"We might have to isolate Domo from Panda tonight," she says. "She might attack him if she smells blood."

I can see that Susan and Anthony have their hands full, so I turn to Sally, who tells me that I've missed only a few weighings so far, and that the first four pigs have made weight, but barely—for some reason, they weighed in dramatically lighter here than they did just the night before on Sally's scale. Sally says that Jackie—the girl whose pig was worryingly light at the weigh-in that I attended—has done a commendable job of bulking up her pig over the past five days, getting up to just about 213 the night before, only two pounds shy of the cutoff. In years past, Sally explains, the judges often let pigs that were that close squeeze in. But tonight, the pig has come in at a devastating 204—too far off to round up. Jackie's pig doesn't qualify for auction.

Sally says that it's not unheard of for animals—especially pigs—to lose weight the day that they arrive at fair. The stress of being transported and adjusting to a new environment can cause them to shed pounds, or even to stop eating entirely. Some animals are so sensitive that the taste of water that's different from what they're used to upsets their stomach. Even so, it's very unusual for a pig this size to lose nine pounds overnight. Sally politely asks the weighmaster—a woman who looks to be in her early twenties—whether the scale might be off.

"The scale is balanced," says the weighmaster.

"But these pigs are coming in ten pounds lighter than they were last night," says Sally.

"The scale is balanced."

Sally still looks skeptical, but there's not much she can do. Starting tonight, the fair staff's word counts for more than that of any parent or 4-H leader.

As the rest of the pigs take their turns, the weights continue to come in as much as fifteen pounds less than the previous night's. Only a few animals are too light to qualify, but even for the heavier pigs, every pound counts, because the auction price will be based on this weight.

Parents shake their heads. One mom pleads with the weighmaster, who doesn't budge.

I watch some moms from another group cheer raucously when a pig comes in at the top of the qualifying range. When the next pig from their group approaches the scale, I listen incredulously as one woman yells, without a trace of humor, "Big money, big money!" These parents seem more invested in the weigh-in process than the kids, who are, as promised, sitting on the fences around the pens. Some are better than others at holding it together; I see more than a few meltdowns over the course of weigh-in.

Susan introduces me to Anthony's grandfather, who is hanging out near the pen, and his grandmother, who is watching from the sidelines on her scooter. They're a sweet couple, full of jokes and stories. Both were 4-H leaders when they were younger, and they're eager to tell me about their family's successes in the show ring. I learn from Grandma that Anthony's aunt was one of the youngest kids ever to win master showmanship at this fair.

Eventually, the weighmaster agrees to reweigh a few of the pigs, but no luck for Jackie—her pig comes in even lighter the second time. Jackie's mom, Jamie, just shakes her head. Jackie is nowhere to be found.

In a different corner of the barn, I find Allison and Sydney sitting on the fence of the pygmy goat pens, wearing their signature trendy jeans and stylish cowboy boots, looking much more glamorous (and about ten years older) than the rest of the kids around the barn. Because pygmy goats are not market animals, they don't have to be weighed in; Allison and Sydney are here tonight because, along with Allison's mom, KC, they are staying at the fairgrounds for the duration of the fair, camping out in an RV that belongs to the parents of another goat project member.

I walk with Allison and Sydney over to the trailer area of the fairgrounds. KC is sitting outside the RV in a beach chair, drinking a soda and chatting with a few other parents from neighboring RVs. Everyone seems to know everyone.

Around 9 P.M., when I'm getting ready to leave, I make my way back to Domo's pen, where I find Anthony and Susan. Domo's ear has stopped bleeding, but it still looks raw, and Domo looks exhausted, lying in a corner of the pen. Anthony assures me that Domo will be fine by tomorrow morning, but luckily, the pig show isn't for another two days.

On the day of the goat show, I arrive at the fairgrounds at 4 P.M., about half an hour before the show is to begin. The midway is in full swing, with game

Figure 16. A meat goat show at the Contra Costa County Fair. Photo by Rafael Roy.

hawkers calling out to passing groups of teenagers and grown-ups drinking margaritas out of bong-shaped plastic containers.

Allison and Sydney aren't in the barn, so I head down to the trailer area, where I find everyone next to the RV. The girls are getting ready: putting on their show whites, frantically searching for lost tie pulls and hats, quizzing one another to prepare for the judge's questions. Sydney is the only one who looks ready to go, hair neatly done up in two braids. Though not officially required under the dress code, braids seem to be the show style of choice for 4-H girls with long hair. I can't believe how much younger she and Allison look in their show whites. Their usual dramatic make-up is toned down; Allison explains that judges frown upon heavy eyeliner.

About a minute before the show is supposed to start, KC manages to wrangle everyone and lead them over to the goat pen, where they discover

Figure 17. Allison gets ready to show her pygmy goat at the Contra Costa County Fair. Photo by Rafael Roy.

that some of the goats' collars are missing. She calls Alexis, her older daughter, who is here for the day helping, to get the other collars from the trailer. This is the first time I've seen KC look frazzled.

The goat show is not in the main arena—it's in an unofficial-looking pen under a big oak tree. The waiting area is a strip of grass, with a few bales of alfalfa lying around. There are no bleachers, so spectators have to crowd around the edges. The scene is chaotic, with 4-H'ers trying to keep their goats from snacking on the alfalfa, which can cause bloat if the goats eat too much of it.

The show is divided by type of goat. Meat goats go first, then dairy goats, then the pygmies at the very end—so there's a long wait for Allison and the rest of the Pleasant Hill group. As I stand with Allison, Sydney, and Alyssa, the mood is tense. Even the goats seem keyed up, bucking and straining on their leashes. Choncho lightly bites both Alexis and Allison. "I've never seen that before," says KC.

Then, all of a sudden, Choncho jumps up and lands on Allison's foot. Allison yelps and hops. Alyssa asks if she's okay, and Allison waves her off. She looks okay for a minute or two, but then she begins sobbing, quietly at first, then louder and louder. KC comes over and comforts her.

"Do you want to scratch?" KC asks. Allison nods.

I persuade a soda vendor to fill a plastic bag with ice cubes for Allison's foot. After a few minutes, a fair medic zips over in an electric cart. He feels Allison's toe and determines that it probably isn't broken.

"But she might have split the joint," he says. "We should get her to the hospital. We could call you an ambulance." KC looks at Allison.

"Mom, I decided I don't want to scratch," says Allison.

KC looks at the medic apologetically.

Allison remains on the hay bale, and Alyssa, Kayla, Sydney, and Alexis continue to fret over her. I can tell that she is enjoying the attention. I don't think she's faking her injury—Choncho did come down hard on her foot— but it doesn't hurt to have a cheerleading squad.

Finally it's the pygmy goats' turn in the ring. When Allison's goat's class comes up, Alyssa helps Allison to her feet. She hobbles into the ring. The judge is a young woman with long, loose blonde hair, cowboy boots, and a chunky necklace.

"I'm going to try to have you walk as little as possible," she says to Allison, who sniffles and nods.

The class is seven or eight big, and Rayne, the goat that Allison is showing, is perfectly behaved. KC leans over to me and says, "Rayne looks really good today. Just last week her coat looked awful, and even yesterday it was all matted in some places and fluffy in others, but that goat cleans up real nice for show."

Sure enough, Rayne takes first in her class and champion for all the pygmy goats. Kayla's goat, Butterball, places first for goats less than a year old. Allison slowly makes her way out of the ring, and her friends descend and help her to her alfalfa bale throne, where she watches showmanship classes until it's time for her class, the senior division.

Surprisingly, eleven-year-old Kayla wins first-year showmanship. "You can really tell this young lady has worked with her goat," says the judge. Kayla beams. But Sydney's goat, Axel, is extremely jumpy, and Sydney places toward the bottom of the class.

When senior showmanship comes, Allison limps back up. When she gets into the ring, she's full of confidence, all business. She remembers every detail of showmanship conventions, always stepping on the correct side of Rayne, grinning at the judge. She handles Rayne with quick, businesslike tugs, and Rayne responds cooperatively. When Allison sets her up, squaring her back legs, she stays posed. The judge has the members of the class trade animals—this is something Allison has told me might happen. She handles another girl's much larger goat flawlessly.

The judge has the class walk, stop, and set up several times. She's taking much longer with the senior group and seems to be having a hard time making up her mind. After a few minutes of indecision, she lines up the class and stops to talk to the first girl.

"What do you call this part of the goat?" she asks, gesturing to the goat's shoulders.

"Withers," answers the girl.

She asks each member of the class a different question, moving down the line. When she reaches Allison she thinks for a minute.

"What kind of goat is this?" she says, pointing to another girl's goat.

"Um. Dwarf?" says Allison, tentatively. The answer is Nigerian Dwarf. KC and I both cringe.

When the judge announces the winners of the class, she turns to Allison and says, "This young lady would have had this class hands down. She had the confidence and the skills. She just needs to work on those questions." Allison takes third, returning dejectedly to her alfalfa bale.

The evening does not improve. Rabbit showmanship is next, which means Allison has to limp over to the small-animals barn, possibly the least charming place at the whole fair. The temperature is about 110 degrees, and an ancient standing fan blows the bunny-poop-scented air around ineffectively.

We watch an endless parade of little kids take their bunnies up to the judge (who, KC informs me, is the cute guy who flustered Allison the year before). At around 9 P.M., after three hours of waiting, I am sweating and sneezing from the rabbit smell, and Allison still hasn't shown her bunny. I give up. After wishing Allison luck, I head home.

Early the next morning, I'm back at the barn for the pig show. I wander through the rows of stalls, looking for Anthony. Before I can spot him, I run into KC, Allison, and Sydney. They look exhausted.

"How was the rest of the night?" I ask.

"Oh god," says KC. "You do not want to know."

But I assure her that I do, so she and Allison tell me the whole story. After I left for the evening, Allison went on to take third place in rabbit showmanship. She missed a few easy questions, maybe because of the distractingly dreamy judge, or maybe because her foot was still hurting and it had been a very long day. Alyssa and Kayla went home with their mother, and Alexis left with her boyfriend. KC, Sydney, and Allison returned to the RV, and around 11 P.M., they all collapsed into bed.

A few minutes later, they heard gunshots. First one, then, a minute later, a few more. KC poked her head out of the trailer. Outside, a 4-H dad from the next campsite over was rushing around his RV, gathering up folding chairs and coolers.

He told KC, "Some drunk guys are out there by the train tracks shooting their guns. I'm taking my kids home, and I recommend you do the same."

But KC doesn't have a driver's license—another 4-H family had driven the RV to the fairgrounds—so she and the girls were stuck. No one felt like sleeping. By the time the girls had to wake up at 6 A.M. to feed the animals, everyone had gotten only a few hours of sleep.

KC tells me they'll try to make it to part of the pig show later. For now, they're going back down to the RV to take it easy for a while.

I head toward Panda and Domo's pen to find Anthony. Both pigs look clean and shiny. Domo still has a scar on his ear, but it's healing. Anthony is wearing what looks like a hazmat suit over his whites to keep them clean. He's spraying the pigs down, since it's already getting hot. He is too busy to talk, so I find Sally, who is walking from pen to pen, making sure the members of her pig group are ready for the show.

"I might be biased," says Sally, "but I think swine is the best show to watch." Pigs are more lively and less predictable than other species, she explains. You can never tell what they'll do in the ring.

The rest of the fair seems to agree. Half an hour before the show is scheduled to start, the bleachers by the main ring are already full. I squeeze in next to Anthony's mother, Susan, hoping that she'll explain the parts I don't understand. She points out the rest of the family: Anthony's dad is there with a video camera, and his grandparents are in the first row of the bleachers. She catches me up on the sheep show: Anthony's lamb, Satchel, took last place in her market class, but Anthony took second place in his showmanship class—a good outcome, especially since it was his first time showing a lamb.

The first class is for breeder pigs. Anthony has decided to enter his smaller pig, Panda, the one who will go to auction at another fair later in the summer. Panda is one of only two pigs in the class. She takes first, but the judge remarks that she is shorn too short for his liking. Susan shakes her head.

"Domo's hair is even shorter, plus he has a little sunburn," she says. "There's no way Anthony is making masters."

A short haircut and a mild sunburn are not ideal, but they're also not grounds for disqualification, according to the fair's code of ethics. The distinction between unethical and simply undesirable showing practices was covered at the ethics and quality assurance training class that I attended a few weeks back. It's perfectly acceptable to "show to the animal's advantage," which means emphasizing the best qualities—and sometimes hiding its flaws—in the show ring. If your pig's shoulders are bigger and more muscular than its rump and ham area, for instance, you should try to drive the pig so that the judge has a head-on view instead of a side or back view. Likewise, if your steer has a particularly thin patch of fur, a combover approach is a perfectly ethical choice.

But the instructor warned that the fair strictly forbids certain appearance-enhancing practices. Some seem obviously unsavory—for example, "drenching" (using a tube to force water into your animal's stomach so that it will gain weight quickly) and air injection (using a syringe to pump air under your animal's skin to make it appear bigger). Others are less glaringly wrong: "icing" (applying ice or alcohol to a lamb's back to make it tense up and appear more muscled) is forbidden, as is using the fake steer fur that I saw at Nasco. At some fairs, the use of certain sprays, waxes, and dyes is also prohibited.

When the market classes begin, the first thing I notice is that they're much bigger than the goat classes: twenty or twenty-five pigs in some. The pig-showing technique is very different from the sheep- and goat-showing style. Instead of leading the animal around, the showman uses only a stick to "drive" the pig—which basically means guiding it toward the judge. It's a lot more haphazard than the sheep and goat shows; there is no lining up in neat rows or walking around in a circle. The main objective is to get your pig close to the judge for as long as possible. This can be a challenge, because many pigs are constantly trying to book it to the nearest corner, often squealing loudly.

At the end of the second class, the judge says, "I'm not liking these red welts I'm seeing on some of these pigs." With the stick, you're supposed to use a light touch, but when your pig disobeys, it's tempting to give it a whack. The rule against using the stick too hard has to do with the animal's comfort, but there's another reason not to whack your pig: you could bruise the meat. In fact, you're supposed to hit the pigs only on their fronts, since the backsides house the most valuable cuts. "Drive from the front seat, not from the back seat," the judge intones into the microphone.

The most common style of judging at fairs is called the Danish, or Group, method: instead of comparing the animals in a class to one another, the judge compares each one to a set of predetermined standards. The winner is the animal (or in the case of showmanship, the showing technique) that comes closest to meeting all the standards. Livestock judging is itself such an art form that many colleges have judging teams that compete nationally. In general, fair judges are highly trained—many have degrees in agriculture, and some are themselves breeders. Although the fairs pay them, the fees are modest; most have full-time jobs as well.

At the end of each class, when the judge picks a winner, he explains what he likes about it. He says things like, "I really appreciate how this one is square through the back," or "This one just feels really muscular through the shoulder when you get your hands on it."

But I find myself completely unable to distinguish between a great pig, a mediocre pig, and a downright bad pig, though I've read up on what to look for. My 4-H swine curriculum book says: "The ideal pig is deep at the heart and long sided. It walks and stands wide and is larger than the average pig of the same age. It is well-muscled, showing natural thickness over the top, and has a plump, thick ham. The ideal pig walks free and easy with good slope to its front pasterns. It is nearly level across the top and has a level rump with a high tail setting."[3]

Today's judge, a charismatic guy who teaches agriculture at a community college in nearby Modesto, is extremely thorough, gradually eliminating pigs until he picks a winner. He identifies the top pigs in each group quickly, then sends them to the pens that line the ring. They wait there while he evaluates the rest of the class.

Anthony's class is seventh: fifteen kids, each with pigs between 260 and 264 pounds. Domo cooperates nicely and looks spiffy, though he's a little red—that's the sunburn that Susan told me about. Domo ends up taking fifth; the judge says he isn't "as square as I might have liked through the back." Susan shrugs. "Fifth isn't bad," she says. "But I hope he does better in showmanship."

Pig showmanship classes are organized by age. There are beginners (elementary schoolers), juniors (middle schoolers), and seniors (high schoolers). Even though Anthony has finished eighth grade, he is still considered a junior.

The juniors are a big class—maybe forty kids or more—so the judge splits them into two sections. Anthony is in the first. When the gate opens, Anthony immediately drives Domo over to the judge, giving him a head-on view. As other kids try to coax their pissed-off pigs away from the edges of the pen, Anthony keeps Domo close to the judge the whole time, glancing back and forth between the judge and the pig. He doesn't seem to see the other pigs in the ring, and the strange thing is that neither does Domo. "It is a shame about that sunburn and the haircut," says Susan. "He's doing so well."

The judge pens Anthony and Domo right off—a good sign that he considers Anthony in the top running so far. Along with a handful of other top kids and pigs from this section, they stay in the pen while the second group in the juniors class takes the ring. Anthony motions for water—not for him, but for Domo, who is getting tired from the heat. It is turning out to be a scorchingly sunny day. It is so hot in the ring that some of the pigs simply lie down. Kids and adult helpers prod and slap at them until finally they get up and shuffle off to the side.

Once the judge has penned his top choices from the second section, he calls the penned kids from both sections—about twenty-five in all—back into the ring for a final round. The kids and the pigs look frazzled and sweaty. A girl in a Future Farmers of America uniform tries to cajole her pig out of the corner, making kissing noises and using her stick, and when that fails, slapping the pig. A younger girl can't keep up with her pig, who races around the ring. One of the cardinal rules of showmanship is to play it cool, and running is frowned upon—so if your pig takes off, there's not much you can do.

Comparatively speaking, Anthony and Domo are looking pretty good. Domo is snuffling around in the shavings agreeably, and Anthony, showing no signs of fatigue, consistently drives him near the judge. The judge pens him once again. Susan and I exchange hopeful smiles. I squint to see if I can detect any sign of happiness in Anthony's face, but he is playing this one as close to the vest as ever, and he seems entirely focused on Domo.

After a few more minutes, the judge grabs the microphone, then dismisses the kids in the ring and brings the five penned kids out. Fifth, fourth, and third are announced. It's down to Anthony and a smaller boy. The judge turns to the smaller boy and praises him for keeping calm with a spirited pig. Then he turns to Anthony.

"This young man really impressed me with his seriousness and his control," he says. He pauses. "And he'll be our first-place winner for junior showmanship today. Congratulations, young man."

I whistle and clap. Anthony shakes the judge's hand and pockets his blue ribbon, still all business. By taking first in his class he has made it into the senior showmanship class and, more importantly, large-animal master showmanship. Susan is ecstatic. I doubt that the word "kvelling" is used often at county fair pig shows, but that's exactly what Anthony's grandparents are doing.

We settle back in to watch Anthony compete in senior showmanship. The difference between the senior showmanship class and the junior one is immediately obvious, even to me. The senior showmen are more confident and decisive with their pigs. They drive them gracefully. There is very little of the desperate whacking that I observed in the junior class. These kids, especially the boys, are also older, bigger, and stronger, which gives them a physical advantage—Anthony is still on the short side. Even among the high schoolers, though, Anthony keeps his cool and drives Domo near the judge consistently and easily. He is penned early once more.

After the judge dismisses the bottom half of the class, he has the top ten kids switch pigs, and Anthony gets stuck with one that is much wilder than Domo, and much more interested in running to the corner. Anthony can't manage to get him out. The judge eliminates kids one by one, until only Anthony, another boy, and two girls are left.

"This is about the toughest class I've seen today," says the judge. "So I'm going to have these four kids come up to the microphone and answer a question. I want each of you to tell me how you would explain to someone who didn't know much about agriculture why you think it's important, and I also want you to tell the crowd whether you think agriculture has a future." He motions to the shorter of the two girls. She grips the microphone and takes a deep breath.

"What I would say to someone who didn't know anything about agriculture is that agriculture is good, and I do 4-H because I like working with animals, and it's really fun?" she says. A couple behind me in the stand smirks.

The next girl, tall and blonde, does a little better. "Agriculture is important because it makes our food," she says. "Of course there is a future in agriculture because everyone needs to eat."

Anthony is next. "I like agriculture and raising animals because it's important to my family. Also agriculture is very important for food, so yes, it has a future."

The last is the other boy. "Raising animals teaches you a lot about responsibility, and also the future of agriculture is very important."

Frankly, I am underwhelmed by all four responses. The kids are so good at showing, but they don't seem to have thought through *why* they are doing it. What's more, they have missed an opportunity to educate the fair-going public about the importance of farming—the whole point of showmanship.

Nevertheless, the judge seems impressed. "Most kids couldn't get up here and ad-lib an answer to a question like that, so let's give these kids a round of applause." We clap. "Now I'll award the prizes," he says. The shorter girl takes fourth, and the other boy takes third. It's now between Anthony and the tall girl. "This young man came to us from the junior showmanship class," he says, gesturing to Anthony. "He continued to impress me here, but when we switched the pigs on him, he couldn't quite get that pig out of the corner. That's both good and bad—good because it means he put in a lot of time working with his own pig. That one is almost a push-button pig. But he also doesn't quite have the skills yet to control an animal he doesn't know. He'll be our second-place winner today, and this young lady over here, who just wowed me with her grace and consistency, will be first."

First in his class and second in the class above! Susan, Sally, and I rush over to Domo and Panda's pen so that we can congratulate Anthony when he gets out of the ring. When Anthony emerges, he is grinning. This may be the first time I have ever seen him smile.

"You made it to masters!" I say. "Yay!"

"Yeah, how am I going to learn how to show a horse and a steer?" he says, still grinning.

"You're going to go ask one of these kids to show you," says Susan, who has switched from celebration mode into action. She goes off to pick up the large-animal masters binder, which Anthony will have to study. Fast. Because the pigs are the last large-animal show of the fair, kids who show them are at somewhat of a disadvantage for large-animal masters, which will take place later this afternoon. If Anthony had gotten into masters for sheep, he would have had several days to prepare.

That evening, I arrive back at the barn just before large-animal masters starts. The horse competition is first. Inside a small, fenced-in arena, a horse is tied to a post. The ten masters contestants are lined up by the gate. Most of the kids are much bigger and older than Anthony—high schoolers. I find a spot along the fence next to Susan, who is watching anxiously.

"They only told them the routine twice," she says. She hopes that Anthony will be able to keep track of everything he has to do; he sometimes has trouble remembering sequences of instructions.

"Is there a lot to remember?" I ask.

"Yes," says Susan. "Watch this first girl."

She's right. The girl who goes first confidently leads the horse to the judge, then runs with it in a semicircle. The horse breaks into a trot. Then the girl brings the horse back to the judge, sets up its legs, and walks around the front of the horse as the judge circles. She finishes by backing the horse up a few steps, then turning it and walking it back to the front of the ring.

"He's going to have a really hard time remembering all that," says Susan.

When Anthony's turn comes, he remembers to do everything except back the horse up.

He did pretty well, I think, but Susan still looks worried. We'll never know exactly how he did in the horse competition, since in masters, the judges don't announce a winner for each animal type. Instead, they take notes, tally up the scores, and announce one overall winner at the end of the contest.

After the kids are done with the horses, to save time, they show the rest of the animals in pairs. Anthony's partner is an older girl named Angel, who seems friendly. The two go to each station, walking their animals around the ring, turning them toward the judges, and answering questions.

It's hard to discern how the kids are doing—they're all so poised. A lot of the masters competition, Susan tells me, comes down to luck. You don't have the luxury of working with an animal that you raised and trained, so if it misbehaves, there's not a lot you can do, besides keeping your cool and letting the animal know that you're not getting riled up. How lucky for Anthony, I think, that not getting riled up happens to be his specialty. But he has trouble controlling the goat he's assigned. It keeps rearing up on its hind legs. Susan gives me a nervous smile and shrugs.

His steer is more cooperative, and I'm impressed by how natural he looks with this beast. He scratches its belly with a long stick, a trick that steer showers use to keep their animals calm. Helping at the steer station is the fair queen, a thin, pretty girl in jeans, a sash over her tank top and a tiara on top of her long, wispy braid. I recognize her as a Future Farmers of America kid from some of the earlier shows. She shadows the masters contestants, and if a steer stops, she makes kissing noises and gives its tail a tug to keep it moving.

Later, at the awards ceremony, the masters winner is announced. It's Angel, Anthony's partner. When I talk to Anthony about it, he says, "I bet I probably came in second."

The next day is Sunday, the final day of the fair—auction time. The bleachers are packed, and at 8 A.M. the auctioneer, flanked by the fair queen and several officials, takes the microphone. We stand for the Pledge of Allegiance, then sit for an endless parade of animals. The kids line up in holding pens, awaiting their turn to bring their animals up to the auction block.

Joking with the crowd and gently teasing a kid on occasion, the auctioneer makes the whole day look spontaneous, but it's actually an impressively choreographed event. Each of the 136 animals up for sale today has an invoice attached to it with its name, its owner's name, and an ear tag number. When it's sold, the fair clerks record the selling price. Then, the "runner"—usually a 4-H or Future Farmers of America member—brings the paperwork over to the winning bidder, who must decide what he or she will do with the animal. There are three choices: the bidder can take the animal home live; have it slaughtered at one of three slaughterhouses; or resell it at the going market price to one of the slaughterhouses, which will then sell the meat to its regular clients, such as supermarkets and restaurants. Most of the local businesses that bid on fair animals each year choose resale, because it's the cheapest option. Jenn Chivers, the Contra Costa County Fair's exhibit representative, tells me that only one or two bidders a year choose to take an animal home live.

The auction hums along, the auctioneer moving efficiently from one sale to the next. The selling prices vary considerably; the grand champion hog goes for $12 a pound, and the 255-pound reserve champion for $15 a pound, a gross profit of $3,825 for its owner, a senior in high school. Parents have told me that it's common for grandparents and family friends to give high school graduation gifts in the form of highly inflated auction bids; later that day another senior fetches $7 a pound on a 1,300-pound steer, for an astonishing total of $9,100.

When it's Anthony's turn, there's a quick bidding war between his grandfather and another bidder, and eventually his grandfather clinches the pig for $3.75 per pound, enough to make Anthony a modest net profit. He beams.

CHAPTER TEN

The Alameda County Fair

Two weeks after the Contra Costa County Fair ends and Allison and Anthony are done with their 4-H season, it's time for the Alameda County Fair, where Chloe and Serena will show their goats and Kelly and her brother, Kyle, will show their lambs. The Alameda County Fair is especially exciting for me, since it was where I discovered 4-H last July. It's hard to believe that it's been a year since the afternoon when I wandered transfixed through the Amador Livestock Pavilion. I imagine myself confidently striding through the rows of pens this year, speaking in 4-H jargon with the kids and their parents. "That is a classy-looking lamb," I might say. "Do you think you will make it to master showmanship?"

So excited am I for the fair to begin that on the first day it's open, before the livestock shows start, I drag a few friends to the fairgrounds in the small city of Pleasanton, to get the lay of the land. The city lives up to its wholesome name. There is a commercial area with upscale shops, cafés, and restaurants. In the residential neighborhoods, immaculately kept houses are adorned with manicured lawns and shining SUVs are parked in the driveways. It's about as different from Antioch, home of the Contra Costa County Fair, as a place could be. Seventy years ago, when the whole region was farm country, the two cities probably didn't look so different, but over the years Pleasanton became affluent while Antioch remained mostly working-class. Unlike with Antioch, Pleasanton's wealth carried it through the 2008 real estate bust.

Figure 18. Serena in the show ring at the Alameda County Fair in Pleasanton, California. Photo by Rafael Roy.

The fairs reflect these class dynamics. Alameda's sprawling fairgrounds take up 270 acres—that's more than triple the size of the Contra Costa County Fair's 80-acre grounds. The Alameda fair lasts three weeks, to Contra Costa's four days. There is also much more stuff to buy at the Alameda fair. Shopping stalls sell dream catchers, customized T-shirts, necklaces, and leather bracelets. One building houses a giant infomercial bazaar. My friends and I peruse items including a special towel that supposedly keeps you cool in the heat, luxury pillows, and many different chopping and grilling devices. Outside there are three separate places to buy hot tubs.

Food stands offer standard carnival fare: funnel cake, Italian sausage, cotton candy. We settle on curly fries, which are served to us packed so tightly together that they resemble a brick. For dessert, we briefly consider deep-fried watermelon on a stick but eventually decide on a bag of hot,

The Alameda County Fair

crispy kettle corn. We buy overpriced beer from a stand called Beers of the World. I recall Andy Hawkey, Chloe and Serena's mother, telling me that she and some other parents from the Montclair club always make sure to bring margaritas in water bottles.

After dinner, I take a walk by the livestock pavilion. A few kids are around feeding and watering, but mostly it's dark and quiet, except for the occasional bleat or moo. This is the one part of the fair that seems basically the same as Contra Costa's: the same earthy smells, neat lines of decorated stalls, and dusty ground.

The first several livestock shows at the Alameda fair (and at many bigger fairs around the country) are called breeding sessions; these are for livestock that aren't intended to be sold for meat. This includes animals that are kept on the farm to breed babies, dairy cows and goats that are evaluated mainly on the size and shape of their udders, and wool-producing sheep and goats. Breeding-week shows don't draw as big a crowd as market shows—probably because the animals that are showing will go home with their owners rather than being auctioned off.

But for Chloe and Serena, breeding week is the main event. Montclair 4-H's herd of Nigerian Dwarf goats are technically dairy animals, even though the group doesn't milk them regularly. Chloe is showing Kajsa, the dark brown mother of the new triplets. Serena has tan-and-white Valentina, the goat whose babies, Huckleberry and Finn, she helped deliver.

I arrive at the fairgrounds a little before 9 A.M. The livestock barn is transformed from the night before, full of 4-H and Future Farmers of America kids in their uniforms, parents, and dairy goats set up on wooden stands for their last-minute groomings. Immediately I spot Chloe running to the goat pen with a bottle of Pepto-Bismol.

"Valentina is sick," explains Serena when I get to the pen. "We think it's stress." Serena attended to her until 10:30 last night. Valentina has been

Figure 19. Serena gets a pre-show pep talk from her sister, Chloe, at the Alameda County Fair. Photo by Rafael Roy.

shaking, having diarrhea, and refusing to eat, and Serena is trying to get her to take some molasses for energy.

But Valentina's class starts any minute, so Serena leads the listless goat over to the holding pen to wait. I follow with Chloe, who fills me in on the other contestants.

"There are some girls I really don't want to show against here," she says. She points to a girl with a rhinestone belt. "She's very competitive. And see that one? She always wins everything. Those girls from Abbie and Tassajara are better than me even though they're younger than me."

Serena's class walks their goats into the ring. To me, it looks like Serena is doing well, standing up tall, staying on the right side of the judge, never losing eye contact. Valentina is behaving herself. She might be a little more docile than usual, but she doesn't look ill like she did just a few minutes ago.

The Alameda County Fair

The judge has the girls switch goats and walk them around. The goat that Serena is given is a little jumpy, but Serena maintains her composure. After a few more turns around the ring, the judge announces the winners. Serena takes fourth.

Chloe sighs. "I thought she deserved at least third." When Serena comes out of the ring, she looks disappointed.

"I recommend you go eat a lot of chocolate rapidly," says Chloe.

Next is Chloe's class. She leads in Kajsa, who immediately transforms into the jumpiest, most agitated animal I've seen in a show ring yet. She bucks and strains as Chloe tries to coax her into the line, and when Chloe attempts to set up her legs, she walks in circles. Chloe keeps trying, but Kajsa won't stand still. While the judge inspects the hooves and udders of the other three goats in the class, Chloe struggles. When the judge approaches, Kajsa is still trying to turn around.

"Can I make a suggestion?" says the judge. "Put your hand on her back, and she'll stop trying to spin."

Chloe does what the judge says, but Kajsa continues to wiggle. Chloe tries to pick up Kajsa's hoof, but Kajsa kicks. The judge steps back.

"I'm going to wait until you're in control of your goat," says the judge. Chloe, never losing her cool, smiles and cajoles Kajsa, but nothing works. Finally the judge steps away and tells the girls to switch goats.

Chloe's new goat immediately starts bucking. Kajsa, on the other hand, calms down a little for the girl who takes her. Andy, Serena, and a cluster of other 4-H kids and parents watch silently through the last torturous minutes of the show. Chloe takes fourth place in her class of four. When she leaves the ring, she shrugs.

Still, Chloe is her usual poised and cheerful self, joking self-deprecatingly about how badly the show went. But beneath it all I think I detect some genuine disappointment. It's Chloe's last year in 4-H. Technically she can stay in the club through age nineteen, but since she's going away to college, she won't be able to attend meetings. On top of the disappointment about being rejected from her top-choice colleges, this underwhelming finale of her 4-H goat career must be difficult.

Next up is "conformation," where goats are judged by breed on both their udders and overall build. The classes go by pretty quickly—Nubians, La Manchas, and Nigerians. No one expects the Montclair herd to do well in conformation; usually the kids from farm country sweep this part of the competition. But to everyone's surprise, Kajsa ends up taking first in her class, then champion Nigerian. The Montclair kids and parents clap and cheer. Andy looks relieved.

The goat show is over around noon. Chloe and Serena immediately start talking about the event they are most excited about: Rabbit Bowl.

When I arrive at the small-animals barn a half hour before Rabbit Bowl is scheduled to start, spectators are already gathering outside the picnic tables where it will take place. The girls are about to go in search of a quiet spot to spend a few more minutes with their rabbit textbooks.

I follow Chloe, Serena, and their two teammates, who set up their study area on a bench inside the Food and Fiber Arts building, in front of prize-winning fair-themed cakes with perfect Ferris wheels covered in fondant. Serena produces a pile of flash cards from her purse.

"Name three wool breeds," she says.

"French, English, Giant Angora," says Chloe, without missing a beat.

"Where is a triangle frown?" asks Serena.

"Behind the ears," answer Chloe and another girl in unison.

"You guys, we should really study more diseases," says one girl. "What's the difference between pregnancy toxicosis and kotemia?"

"They're the same thing," says another girl. "That's a trick question."

"What's a bacterial infection of the eye due to staph?"

"Weepy eye."

"How long can your rabbit be quarantined?"

"Three weeks."

The studying goes on like this until it's time to head back to the small-animals barn, where the crowd has grown to about a hundred kids and parents. The junior teams—elementary and middle schoolers—go first. Montclair wins handily.

Chloe and Serena hug the younger girls. Andy pours me a homemade margarita in a paper cup, which I raise with a few of the proud parents. The seniors—Chloe and Serena's team—are up next.

Rabbit Bowl is a round-robin-style competition, and Montclair is toward the end of the first round, so we watch the other teams. The questions for the seniors are technical: Define malocclusion. What is ear canker? What is the ideal length of a Rex or Mini Rex coat?

Montclair's turn finally comes. Their opponents are three girls from another club wearing superhero costumes—blue suits and red sequined capes and headbands.

The first question is for Montclair: What are the varieties of the Silver? No one knows the answer, so it goes to the other team. "Gray, Brown, and Fawn," says one of the superhero girls. Correct.

"Describe Jersey Wooly wool," says the judge. Montclair is silent.

Another one of the superhero girls buzzes in. "The texture is very coarse." Correct.

"Name five markings on the Dutch that are allotted points."

A superhero girl lists a few, then falters. This is Montclair's chance to pull ahead.

But Chloe answers incorrectly, too.

During the rest of the competition, Montclair gets only a few correct answers. For the first time since they started competing years ago, they don't take first in Rabbit Bowl—or even second. At the end of the competition, they end up in seventh place.

"It was so hard!" wails Chloe afterward. "We're supposed to be good at Rabbit Bowl and showmanship, and we're supposed to lose in the breeding competitions. It was totally the opposite this year."

"There were a lot of questions about breeds," says Serena, shaking her head. "We studied more about diseases."

Later, when Chloe and Serena have disappeared, I catch up with Andy, who is cleaning up her cooler and getting ready to go home.

"Do you think Chloe and Serena were upset about Rabbit Bowl?"

"I think they were just so surprised. And I do think that it's a rotten last Rabbit Bowl for Chloe. I wish she could have finished on a better note."

When breeding week ends and market week begins, it's as if someone has turned up the volume in the barn by about five notches. There are more kids in show whites, more parents carrying cameras and grooming supplies, and bigger crowds on the bleachers in the arena. Even the animals are making more noise.

I take a seat next to Kelly's mom, Teri, who promises to give me all the dirt on the lamb market competition today so I'll know what to look for in showmanship tomorrow.

Group after group of kids come into the ring, lead their sheep around, and file out. The judge, a stocky man with a buzz cut, keeps things moving. He seems to be handling the sheep a lot, patting each one's back before making his decision. Teri explains that he's trying to tell how "finished" they are—the ratio of muscle to fat. The layer of fat should be about the width of a penny.

Sheep classes tend to be big, and ideally you want the judge to notice you right away. Teri says that Kelly always tries to be one of the first to file into the ring, and when the judge lines the class up, sometimes she stands slightly forward so that when the judge looks down the line of lambs he'll see her first.

Teri also points out Kelly's main competitors. There's a family of three girls whose last name, Castello, I recognize from the bulletin board where prizes are announced. They show sheep, pigs, goats, and steers, all with astonishing success. The oldest one, Alex, is Kelly's age. There's also a pair of twins whom Teri describes as "nice girls and frighteningly good at showing."

I'm finally starting to pick up the language of sheep showing. If the judge says your lamb is "rugged," "angular," "square," or "fresh," you're doing well. You'd never want your lamb described as "stale," which means not old (as I'd originally thought) but lacking in muscle.

Kelly has a few lambs in this show; one takes eighth, and a few classes later another takes fourth. Finally, toward the end, Kelly shows Walter, one of her favorite lambs. The judge seems to be deciding between Walter and another lamb for first place. Of Walter, he says, "This one is more rugged in his general appearance than the first, but he's a little plainer than the first." Walter takes second. Kelly looks pleased.

By the time I arrive in the livestock pavilion on showmanship day, around 8 A.M., Kelly has already been here for hours. I find her putting the finishing grooming touches on several lambs. Because 4-H rules state that kids can help other kids, but parents aren't allowed to touch the animals, four of Kelly's friends from school—kids who aren't in 4-H—are helping Kelly at the fair today, handing her brushes and clippers when she needs them, holding sheep on leads while she is in the ring. Because Kelly's 4-H friends will be busy with their own animals, her school friends will be indispensable.

I leave Kelly to her work and follow Teri to the ring, where the beginners' showmanship class—an array of nine- and ten-year-olds—is just getting started. I recognize Carly, the youngest member of the formidable Castello clan, younger sister of Alex and Courtney. She stands out among the beginners for her focus and her size—she's one of the littlest there.

The first time the judge asks the kids to brace the lambs, Teri leans over to me and whispers, "Watch Carly when she braces her sheep. Her stance is going to be much lower than the other kids'."

Sure enough, I see that Carly's legs are wider apart, and her center of gravity is lower. She's almost in a split.

"That looks hard," I say.

"It is," says Teri. "That's the Castello girls' style of showing. It's very low and requires a lot of leg muscle. You have to practice a lot to do that. Some judges like it, and some don't."

"Does Kelly ever show that way?"

"She doesn't like to, but if she sees that this judge prefers the low showing style, she might try to get lower out there."

The kids continue walking and bracing. Each time, Carly's stance seems lower. Carly's little black lamb is one of the jumpiest in the group. A few times, he strains so hard that he pulls Carly to the ground. When the judge lines the kids up, he places Carly third from the front.

"He can't be considering her for third place," I whisper to Teri. "Her lamb keeps running away."

Carly stares intently at the judge, and her sheep bucks and pulls her over again. She struggles to her feet and gets her sheep back into the low brace, only to have it pull away once more.

The judge walks slowly and deliberately around to each competitor, asking questions. There's a trio of girls at the front of the line that he seems to be taking extra time with. They have good control of their lambs, great eye contact, and self-confidence. Based on the shows I've seen so far, I figure that he's considering them for first, second, and third place.

Finally he takes the microphone and faces the crowd. "It's hard for anyone to show a sheep, but it's even harder to show a sheep if you're this size," he says. Then he gestures to Carly. "This young lady's brace technique is the most appealing I've seen in this class, but because of the control issue, she'll be in second place." One of the girls from the front of the line takes first.

I'm floored. Carly's lamb was easily the worst behaved in the class—and yet she has still placed. Teri raises her eyebrows at me and points over to the side of the ring, where Kelly and a few other older 4-H kids are observing.

"They're watching all this. Now they know that this judge wants everyone to be really low. None of those kids usually show like that. It'll be interesting to see whether they change their technique to appeal to this judge."

Next is the junior showmanship class, Kyle's division. Because Carly took second in beginner showmanship, she moves up to compete in this age division as well. The judge quickly sorts the class of twenty-seven into two groups, one of which is composed of the best showmen. He places Kyle, Carly, and Carly's older sister Courtney—whose stance is as low as her little sister's—toward the front of the line. Carly's sheep bolts again.

Kyle's brace style isn't as low as the Castello girls', but he makes the whole thing look easy. While the other kids anxiously search the judge's face for any response to their performance, Kyle's attitude toward the judge is friendly but relaxed. He gives the impression that he is in total control of his sheep without even trying.

The judge has the class take their sheep around a few more times, consults with each one individually, and then picks up the microphone. Carly takes eighth, and the judge places the other kids in the top group in seventh through third. Only Courtney and Kyle are left. "Ultimately the best showman in this competition has what I would call 'the eye of the tiger,'" says the judge. "Also excellent form." After a dramatic pause, he turns to Courtney and shakes her hand. The crowd cheers. Kyle takes second, and Teri cheers, but then she turns to me.

"This is not good," she says. "The next two classes are the most competitive ones. Everyone's going to be showing like that. This is not Kelly's strong suit."

I walk over to where a few older girls in whites are watching. Kelly is not among them. I'm guessing she's back by the pens, preparing. I listen to the girls' conversation as the intermediate class—ninth and tenth graders— begins. The class braces for the first time, and one of the girls on the sidelines whispers to another, "Look at her. She is so low."

"Fuck that," another girl replies. "I am not going to go out there and do the splits. That's not how I show."

"This is my last year," says another one. "I am not going to let some judge come in and tell me I've been bracing wrong for nine years."

Predictably, Alex, the oldest of the three Castello girls, takes first, and another girl with a very low brace takes second. I have to admit that even without their unique style of showing, the Castello girls stand out. All three have a real queen bee vibe about them. If it's possible to look glamorous while bracing a sheep in a 4-H uniform, these girls have done it.

Since Alex won in the junior class, she has moved up to compete against the eleventh and twelfth graders in the senior division—Kelly's class. The senior group is composed of eighteen kids, only three of whom are boys. I recognize some of the kids from earlier shows: the twins that Teri pointed out to me and one other girl from Palomares 4-H, Kelly's club.

The class walks their sheep for a while, and then the judge stops every- one for the first brace. The low brace style seems to have spread to about half the class, including a few of the girls I overheard pledging they wouldn't do it. Kelly, I notice, is not among them. She's all focus and poise, but her brace is as tall and sturdy as ever.

The judge eliminates about half the group from the top running. Alex, Kelly, the twins, and the other girl from Palomares all make the cut. The judge has the top contenders walk, brace, walk, and brace some more. To me, Kelly's performance seems flawless. Even when the judge has his back turned and is giving feedback to the other kids, Kelly keeps her sheep per- fectly set up.

The judge talks to each of the class members individually and then addresses the crowd. "These are top-notch showmen," he says. "A few of these kids could be competitive at any sheep show in the nation. The ones who will place top in this class have all nailed the transition from the walk to the brace, which is the most challenging part of the sheep show. It should be graceful, even if that sheep is struggling. It should not look like a scene from Dancing with the Sheep." The crowd chuckles.

"The question I asked them was an important one," he says. "What type of digestive system does a sheep have? What makes that unique from a human or hog? Some answered right, some did not."

Teri whispers to me, "Kelly knows that one!"

Finally, he calls out the winners. Alex is first. One of the twins is second, a girl I don't know is third, and the other twin is fourth. Kelly is sixth. Of Alex, the judge says, "There are so many little things that she does with her showmanship that can be imitated. She understands the little as well as the big." I tell Teri that I think the same is true of Kelly. She nods.

After the show, Teri and I run over to Kelly. I think she's going to be upset, but she's actually looking relaxed for the first time in days. Most of the girls in the class are smiling and hugging.

"You were so low out there!" Kelly says to one of the twins, laughing.

I can't believe how cavalier Kelly and her friends are being about this judge, who clearly doesn't know what he's talking about. What is the point of working all year when you might end up with a judge who requires a completely different style of sheep showing? Who *is* this guy, and why *does* he want kids to squat with their sheep?

I decide to find out. From a quick Google search, I learn that his name is Kolby Burch and that he runs a sheep-breeding operation in Iowa. As I dial his number, I am nervous. I don't know why. It's not like he's going to berate *me* for not being low enough. But after a nail-biting sheep show, I'm scared of this guy.

So I'm surprised when he's friendly and forthcoming over the phone, not at all like the stern arbiter he was in the ring. Born and raised on a cattle and sheep ranch in Wyoming, Burch was on competitive showing and judging teams and winning state championships by the time he was in high school. He continued showing and judging competitively in college and went on to receive a masters degree in agriculture from Iowa State. He is now a professional judge, working thirty or forty shows a year. It's not lucrative; after expenses he makes $200 or $300 per show. His bread and butter is back at home in Iowa, where he raises show sheep, teaches agriculture classes, and serves as the Future Farmers of America advisor at a high school.

He tells me that he wants to inspire in the kids a love of livestock. "I've probably never been somewhere without looking at livestock, which might sound crazy," he says, laughing. "But when I went to Scotland and Belgium, I went to a livestock show there. When I went to Canada, I toured cattle ranches. Most of my friends are people I've met at livestock shows."

I ask him what he thought about the showmanship competition that he judged at the Alameda County Fair.

"Some of those kids were getting into the low style," he says. "I think if you don't get your center of balance right, you end up pulling on the head and the neck. The style that I was pushing makes sure the animal is comfortable—they're propped up against the center of your body. You have more surface area for the animal to push against."

"Is the low style the key to winning a show that you're judging?"

"Not at all," he says. "I look for someone who is self-assured. They need to be aware of their surroundings. I've been around this enough to know an intelligent showman. When we have them on the brace for side profile and the lamb balks a little bit, if the kiddo takes their right hand and reprops them, that tells me that they're an intelligent showman."

Figure 20. The auctioneer handles bidding at the Alameda County Fair. Photo by Rafael Roy.

"What about sore losers?" I ask. "Does anyone ever have a fit or accuse you of being unfair?"

"Not really," he says. "You have to have a thick skin if you're going to do this. Kids learn that when they're real young. The ones who can't bear to lose, they usually decide that showing is not their thing."

This explains a lot about the good sportsmanship I've observed at the fairs. I saw a few ringside tears, but mostly they were from younger kids. The teenagers had learned how to lose gracefully.

The fair is a strange phenomenon. For months before it happens, it's all that anyone can talk about. Then, when it finally arrives, time speeds up. It's dawn in the barn, and kids in jeans are sleepily carrying water to their animals. Then it's noon and someone's missing his tie, and someone else is trying to blow-dry three sheep at once, and someone else is worrying over a pig who is too stressed to eat. The shows pass in a blur of hugs and tears and music wafting over from the midway, and then the barn empties, and it's chore time again, and in just a few short hours it all starts again.

On the day of the Alameda County Fair auction, I wake up early, feeling sentimental. For me, this is the last day of fair season and also my final day in the Amador Livestock Pavilion. But the mood at the fairgrounds is jubilant. The biggest crowd I've seen yet has squeezed into the bleachers, though the auction won't officially start for another forty minutes. The auctioneer's podium is set up. Some kids are already in holding pens with their animals, and others are racing around in their whites.

I scan for familiar faces, and almost immediately I spot one: a Montclair mom whom I've sat with at shows a few times. She tells me that Chloe won advanced rabbit showmanship. After the disappointment of goat showmanship and Rabbit Bowl, Chloe's last year as a 4-H'er still turned up one big success.

The auction begins. One by one, kids in their whites take their animals up to the podium. Since Chloe and Serena don't have any market animals, that leaves only Kelly, who is selling two of her lambs, the maximum number allowed at auction. The one that did well in the market class goes for an impressive $9 a pound to Kelly's neighbor down the road, and Kelly's grandmother buys the other one for $8 a pound. After paying her dad back for the lambs she bought from a breeder, and for feed, shearing, and the 5 percent auction fee, Kelly ends up making $859. She'll use some of her earnings to buy lambs next year, and the rest will go into her college fund.

It's scorching hot, and the barn is getting steamy. This auction will last for several more hours; there are about four hundred animals in total, more than twice as many as at the Contra Costa County Fair's auction. I decide to call it a day. I take one final look around and absorb the scene: the auctioneer chanting over the animal noises, the kids waiting in the pens with the animals that they raised, the parents and friends packed into the bleachers.

In just a few hours, everyone will go home, put the fair out of their minds, and get on with their lives, while the barn will remain empty all fall, winter, and spring.

CHAPTER ELEVEN

"It's in My Blood"

A few weeks after the fair, I pay a visit to the Castello sisters, hoping to figure out how they became so good at showing. As I drive out to the family's cattle ranch, I remind myself that these girls are just 4-H kids, but I can't shake the feeling that I'm going to meet celebrities—the rock stars of Alameda County Fair.

The Castellos live in the unincorporated community of Mountain House, on the eastern edge of Alameda County. Farms dot the flat landscape, and cows graze on either side of the country road that I turn onto from the freeway. When I arrive, Alex and her mom, Elizabeth, are waiting outside their ranch house, the pastures unfurling behind them.

Elizabeth explains that this ninety-six-acre cattle ranch has been in her husband's family for two generations. She and her husband both grew up showing livestock. Her daughters learned to brace sheep almost as soon as they learned to walk, and they haven't stopped since. During jackpot season, she takes all three girls to shows almost every weekend.

"Alex, go inside and get your sisters," says Elizabeth. Alex runs inside and returns a few seconds later flanked by Courtney, thirteen, and Carly, ten. All three are in T-shirts, jeans, and boots.

"Go ahead and show her your buckles," says Elizabeth.

The girls lead me over to a folding table groaning with 4-H championship belt buckles.

"So you win these for lambs?" I ask.

"Actually, we show everything," says Alex. "Sheep, goats, steer, and swine."

The girls take me to see some of their animals. We pass three sheepdogs, then a sizable goat pen, then an even bigger sheep pen. I request a sheep-showing demonstration, so Alex takes a lamb out of the pen and braces it.

"I notice that all of you have a very specific way of showing sheep," I say. Carly giggles. It's clearly something they've heard before. "The judge that I saw really, really liked it. And I wanted to ask you guys how you learned to do it."

"I dunno," says Alex, smiling. "It's natural?"

"It's natural?" I repeat, raising my eyebrows.

"We've just been doing it for so long," says Alex.

"It's, like, in my blood!" says Carly, giggling.

"We were taught by my mom, I guess," says Alex. "It's basically like squatting for at least ten minutes at a time. It hurts."

As the girls put the lamb back in its pen, I say, "You know, a lot of the other 4-H'ers I've been talking to live in the suburbs or the city. Have you guys ever thought about what it might be like to live there instead of here?"

"Oh, this is way better," says Courtney.

"Way better!" echoes Carly. Gesturing to the pastures, she says, "You just go out there and do whatever you want to do." She thinks for a minute, then furrows her little brow and says seriously, "In the city what I'm worried about is all the gangs."

We all crack up, and Carly looks indignant. "No, really," she protests.

"I like it out here because you don't feel crammed with other houses," says Alex. "I wouldn't be able to live with people next door to me."

"Or traffic," adds Carly. "Beep beep beep."

When the sun starts to set I leave Alex, Courtney, and Carly to do their mucking and feeding. Even though I have all but abandoned my quest for authentic 4-H'ers, I can't help but notice that these three meet most of my original criteria. Live on a farm? Check. Animals at home? Check. Cowboy boots? Check.

But somehow finding them feels anticlimactic, as if they've taken the magic out of my original vision. There is no trick to their 4-H success. The Castello girls' real advantage at the fair probably has less to do with their low showing style than with their upbringing: a hundred-acre working ranch to watch and parents who taught them to show livestock from the time they were tiny and brought them to shows every weekend. Carly was hamming it up when she told me that showing sheep was "in her blood," but to some extent she was probably right.

The Castello girls were probably the children *least* in need of a 4-H agriculture education. Unlike Lilly, the little girl who thought chocolate milk came from brown cows, they lived on a working ranch. Even without 4-H, they would have grown up understanding where their food came from and why farms were important. Meanwhile, most non-rural 4-H clubs in the United States focus on projects that aren't directly related to agriculture: rocketry, robotics, filmmaking.

So has 4-H neglected to teach its urban members about the importance of farms? Not entirely. Since the 1960s, 4-H has increased its agriculture education in cities through after-school programs, classroom-based curricula, and camps. Nevertheless, there is still a divide between rural 4-H'ers and their urban and suburban counterparts. At times, the 4-H leaders I spoke to seemed conflicted as to whether or not agriculture education is still part of the organization's mission. Livestock projects are still the most popular 4-H activities in farming communities—and yet, when I asked National 4-H Council staffers about the club's philosophy on livestock projects, many emphasized that animal raising is a very small part of the contemporary 4-H program, and that they wanted to fight the perception that 4-H was a "rural thing." I was told that 4-H does not keep track of how many of its alumni pursue careers in agriculture. The point of 4-H, one leader said, was not to teach kids about farming; it was "youth development." That leader noted that 4-H's mission statement—"4-H empowers youth to reach their full potential, working and learning in partnership with caring adults"—doesn't mention agriculture, nor does the "about" page on the National 4-H Council's website. It doesn't even name agriculture among its areas of focus; instead it lists "science, citizenship, and healthy living."

By keeping the "rural thing" separate from the rest of 4-H's urban programs, leaders may be unintentionally widening the gap between farm-country members and the rest of the organization. One rural 4-H leader that I spoke to said that she resented having to prepare her club members to answer "ignorant questions" at the fair. "People from the city assume they know what 4-H is all about, but most of them frankly have no idea," she said. That attitude could explain why the finalists in the pig showmanship competition struggled to answer the judge's question about the future of farming: maybe no one had ever pushed them to explain the connection between their work in the show ring and our food system.

Some of the 4-H curricula reinforces the idea that city people are hopelessly ignorant about farming. In a 4-H hog project curriculum workbook titled *Going Whole Hog*, a section called "Thinking of the Neighbors" asks 4-H'ers to draft responses to various complaints.[1] One hypothetical scenario is described as follows:

> Your family owns a 200-sow herd and farms 750 acres of crop land about 1.5 miles from town. Several of your neighbors who used to be farmers have recently sold off 5 acre lots to employees of a new computer company in town. These new employees want "the country experience."
>
> When the wind blows from the east, you can usually count on getting complaints about odor from these people. One of your new neighbors is a lawyer and an environmental activist. He has just stopped by your house to complain about the odor. What should you say? Be sure to reply in a way that acknowledges the legitimacy of his concern on some level.

The passage has a clear point of view. The environmental activist lawyer is a pampered city slicker who wants "the country experience"—but none of the unpleasantness that's part of it. The scenario reflects a dynamic that is much bigger than just 4-H: many farmers feel misunderstood by the non-farming public. A 2011 survey by the US Farmers and Ranchers Alliance found that 86 percent of farmers and ranchers believed that the average consumer knew little or nothing about modern agriculture.[2] More than half thought that Americans had a "completely inaccurate perception of farming and ranching." Four-fifths believed that consumers "have little to no knowledge about proper care of livestock or poultry."

It's not just the farmers and ranchers who think that city folks don't understand anything about their world. Remember that 2010 Oregon State University survey—the one that quoted a teacher saying that all she knew about farms was "what I've seen on TV, *Little House on the Prairie* like"?[3]

When I read that bit of the survey, I scoffed at that teacher's ignorance. But then again, I had spent months trekking all over California in pursuit of a wholesome, pigtailed *real* 4-H'er. Yes, it was possible, I admitted to myself, that I, too, harbored a few farm stereotypes.

Because of its geographically diverse membership, 4-H is well placed to teach city folks about their farm-country counterparts: about 70 percent of 4-H'ers live on farms, in rural areas, or in small towns and their suburbs.[4] So why doesn't 4-H facilitate more meaningful interactions between its urban members and agricultural communities?

One program in Oregon is doing just that. In 2005, a group of middle-school students from Portland joined ranchers and environmentalists at a hearing about a controversial wolf reintroduction program in the eastern part of their state. The students recited a poem that they had written from the points of view of a wolf, a rancher, and an environmentalist. It ended with the rancher describing his feelings of guilt about shooting a wolf. The poem raised some eyebrows among ranchers, many of whom thought that

"It's in My Blood"

the school had told the children only one side of the story. The *Oregonian* newspaper ran an editorial by a rancher called "Git Along, Lil' Liberals"; another newspaper published an article titled "Student Testimony Teaches a Lesson in Irresponsibility."[5]

Maureen Hosty, a 4-H staffer in Oregon's Multnomah County, read the reactions and found the situation troubling. For years she had seen fissures between Oregon's ranch communities and its cities, and it bothered her that the divide was persisting into the next generation. She was sure that the two sides had common ground. But they rarely interacted.

Then she came up with an idea: wouldn't it be neat if urban kids and ranch kids were given the chance to switch places? A few months later, Oregon's 4-H Urban/Rural Exchange was born. In 2005, twenty seventh graders and five adults from the Sunnyside Environmental School in Portland went to live with ranch families in Grant County, Oregon, a five-hour drive east over the Cascade Mountains from Portland. For six days, the Portland students participated in every aspect of the ranchers' lives, from feeding and caring for cattle to attending a local school to going to church on Sunday. The program has taken place every year since.

In 2012, Oregon Public Broadcasting made a short documentary about one year's exchange.[6] In the film, both the ranchers and the students describe feeling nervous beforehand. "We figured they were brainwashed, that they were dyed-in-the-wool environmental activists," says one rancher. "That has basically been the ranchers' contact with people in the city: the ones who make the most noise." One of the Portland students recalls that the first thing she noticed after getting off the van in Grant County was cowboy hats. "We laughed, because we were like, they really do wear cowboy hats?"

But the documentary shows that over the course of the trip, the Portland students got to know their hosts and the ranch lifestyle. Sometimes, the interactions were awkward. The Portland kids, whose school religiously recycles and composts, were shocked to see only garbage cans at the school they visited in Grant County. "There's so much stuff they could have recycled, and they threw it away. . . . We said they were slapping Mother Nature in the face," one student remembers, laughing. "We were just joking around, but I don't think they thought it was very funny." She continues: "Their definition of environmentalist . . . is someone who causes trouble, so we have to kind of switch it around to make us not seem like the bad guys."

Hosty says that the kids' perceptions of the ranchers often change dramatically during the course of the visit. "The Portland kids tend to think that the ranchers lead a simple lifestyle, that they might have hay sticking out of their mouths," she says. "Then they see that they have to be a vet and a soil scientist and a car mechanic and everything else. They see how multitalented ranchers are."

The city kids are also impressed by the ranchers' love for their livestock. Before the visit, "they think the ranchers will just look at their cows as a product," Hosty says. But when they get to eastern Oregon, "they can't believe how hard the ranchers work to take care of their animals. They put their animals before themselves." A few of the kids who were vegetarian before their visits even began to eat meat after seeing how well the ranchers treated their livestock.

"What was most rewarding to me was when we had the debriefing the first year," says one rancher. "Several of them said, 'We just thought you turned the cows out and made lots of money,' and I thought, 'It's good that they learned that there's an effort to it.'"

Likewise, the kids from eastern Oregon learn during their visit to Portland that life in the city isn't exactly how they have pictured it. Some are surprised to see that their host families don't live in high-rises. "They assume that people in Portland wouldn't know anyone else because the city is so big," says Hosty. "They are surprised to see that we have these things called block parties, or that families in a neighborhood will get together for a soup night."

Many of the kids and their host families have stayed in touch for years; one informal survey found that 40 percent of the Portland students had been back to eastern Oregon to visit the ranchers they met, and most had taken their families with them. One year, after an exchange, about thirty families from the Portland group decided to buy beef directly from one of the host ranchers.

As far as Hosty knows, Oregon's Urban/Rural Exchange is the only 4-H program of its kind. It would be easy to replicate; the only expense is the cost of transporting the kids between Portland and eastern Oregon—about $150 per student, which is paid by the children's families or scholarships from Oregon's 4-H foundation. "I don't know why more 4-H programs don't do this," Hosty says. "That one-on-one time is so important. Sitting down at the table with the rancher and his wife, or going to check on the cows in the morning. You can't get that just from reading about it."

Another bright spot in urban 4-H agriculture education is in Austin, Texas. In 1992, Austin 4-H leaders noticed that city kids seldom had an opportunity to visit a farm. So they decided to bring farm experiences to the city. Today, Austin's 4-H Capital Urban Animal Science program allows third through fifth graders to raise show goats at public elementary schools in Austin. Founded with the help of a grant from the USDA, the program was originally part of a push to provide more after-school activities for high-risk children. It still targets schools where many kids receive a free lunch. In 2012, the 4-H Urban Animal Science program ran sixty goat projects at nine different schools.

To Lydia Domaruk, the Texas A&M county Extension agent who runs the program, agriculture education is a central goal. "We really want [the students] to understand where their food comes from," she says. "I have seen the kids make the connection that this is an animal, and that animals are a big part of our daily lives."

Austin's 4-H school goat program is expensive—running $22,000 to $26,000 a year, not including staff salaries. But the 4-H staffers found a way to pay for most of the costs through a partnership with the Austin Independent School District and Texas's federally funded statewide after-school initiative. Over the past few years, because of the success of the program, it has expanded, and the county now chips in as well, buying the goats at the beginning of each school year. Seven of the nine participating schools also contribute some money.

On a hot afternoon in October, I visit the after-school 4-H goat program at Galindo Elementary School. Fifteen children and a few chickens mingle in a grassy yard behind the school. Inside a large, simply constructed pen, seven goats begin to prance around, visibly eager to greet the children. The kids and the animals don't know each other well yet—the goats arrived at Galindo just two weeks earlier. They're still jumpy and shy, so today the students will "just try to bond with the goats and handle them," Domaruk tells me. It will be another few weeks before the goats will trust the children enough to begin training for showmanship.

The students divide up into pairs, and each pair is assigned a goat to catch in the pen. Once the children have caught the goats and attached their harnesses, they fasten the leashes to a fence on the inside of the pen. Most of the goats and some of the children look very nervous. But one chubby little girl in a neon pink T-shirt bearing the slogan "Girl Power" looks right at home.

"Get him closer to the fence," she instructs her partner, a smaller boy with a buzz cut. "That will make him feel safer. You have to put your knee in front of him. That will make him calm down." The boy halfheartedly offers his knee to the goat, who backs up, straining against his leash.

"Not like that, like this," says the girl exasperatedly, bracing the goat, just like I had seen Kelly Semonsen do with her sheep. The goat did seem to relax, leaning into her a little. The girl must have learned to brace during last year's program. She reminded me of Charlie Brown's friend Peppermint Patty.

Farther down the fence, another pair of kids look worriedly at their goat, who is lying down.

"Is that goat passed out?" asks a little boy, a note of panic in his voice.

"Just pet it some more," suggests another boy.

"Can you get sick from touching a sick goat?" the first boy wonders.

"Great question," says Domaruk. "It depends on the sickness, but not usually. These goats are vaccinated against a lot of diseases."

Another boy announces to the group that his goat has nibbled on his pants. The excitement continues until it is time to unleash the goats from the fence and leave the pen. The first boy out of the enclosure says he wants to check the school's chicken coop for eggs, and then everyone else wants to join him.

"How many? How many?" yells Peppermint Patty.

"Five," says the boy. "No, seven! Seven eggs!"

"Did you know we sometimes put fake eggs under the chickens to help them lay real eggs?" Peppermint Patty asks me as we crowd around the coop. Then she squeezes past the boy who found the eggs and picks one up. "I want to take an egg home," she says. "My mom has never seen a chicken, so I want to show her. I'm going to tell her there's a chicken inside, even though there's not. Then I'll say, 'Just kidding.'"

A few months after my visit to Galindo Elementary, I arrange for a phone call with a Galindo goat-project member: eight-year-old Samantha Stoltje. A third grader, Samantha has been in the program for only four months, but she tells me she has already learned a lot about goats: how to feed them and give them fresh water every day, and why it's important to make sure their food doesn't get wet and moldy.

"Outside of 4-H, what do you think goats are used for?" I ask.

"Sometimes they use fainting goats to protect their other more expensive goats," says Samantha. "When some animal comes to try to eat the goats, the fainting goat will faint, and the animal will take that one and not the other, stronger ones."

I probe further. "So why would they want to protect the other goats but not the fainting goats?"

"Maybe the other goats have more meat."

I ask her how a farmer might tell which goats have the most meat.

"They probably have the goat stand up straight with its legs apart, then they feel around the back and the legs to see if it has enough muscles and meat, like judges do at a goat show. He would want a strong goat with lots of meat."

"Why would it be good to have a lot of meat?"

"So if you wanted to auction them off, someone will pay for a good goat with good meat."

We talk a little bit about other reasons that goats might be valuable. I tell her that I am thinking of a way in which cows and goats are similar, and I wonder if she can guess what I am thinking. She answers that both cows and goats give milk.

I expected Samantha to talk mostly about her own experience with goats, but she has proven entirely capable of answering my questions about uses of goats and the business of goat farming. It's possible that Samantha has picked up this knowledge outside of 4-H, but I think that's unlikely; toward the end of our conversation I ask Samantha if she has ever been to a farm. She tells me that she doesn't think so.

Evaluations suggest that Samantha is not the only participant in Austin's 4-H school goat program who is learning. In 2012, at the beginning and end of the project, sixty-four children took a test about goat anatomy, care, and breeding. Domaruk shared a copy of the test with me. I thought some of the questions were quite challenging.[7] For example:

What percent protein is 4-H Capital's goat feed right before the show?

    a. 10%
    b. 16%

c. 22%

d. 18%

Some goat breeds have very valuable hair. Which of the following are *not* goat hair types?

a. Cashmere

b. Pashmina

c. Mohair

d. Wool[8]

After completing the program, 82 percent of the participants increased their scores on the test by 20 percent or more. At the county fair, one member of the program took reserve grand champion in junior showmanship. Another member took third place.[9]

The city kids in Austin's school goat groups and Oregon's exchange didn't have agriculture "in their blood," as Carly Castello did—but 4-H leaders had provided them with authentic farm experiences.

I think it was that same spirit that kept me interested in the urban and suburban California crew. Raising animals wasn't easy for the families I had followed. The parents were forever shuttling their children back and forth between home and their animals and making long drives out to feed stores. The kids had to explain again and again to their non-4-H friends at school why they raised lambs instead of playing soccer. But the club communities had decided that the payoff—an agricultural education—was well worth the hard work.

Conclusion

After the Fairs

One evening a few weeks after the Alameda County Fair, I drive up to the Hawkeys' house in the Oakland hills. Chloe is leaving for college tomorrow, and her parents have invited their friends and her friends to gather on the deck for a party. When I arrive, a spread of tomatoes and ricotta, pasta, and homemade bread is arranged on a table. Teenagers and adults cluster around, and I recognize a few 4-H kids and parents among them.

Andy gives me a big hug, offers me a glass of wine, and tells me that Chloe has been trying to muster more enthusiasm about the college she chose—Whitman in Walla Walla, Washington, not her first choice. Last week, she signed up for an orientation camping trip in a wilderness area not far from the school. Andy hopes that the trip will help Chloe fall in love with the gorgeous landscape. "I was thinking about Chloe's adjustment to college and how she would deal with being homesick and missing us," she says. "But it wasn't until today that it hit me that she's really leaving." She tears up.

Serena tells me that she is sad about her sister's departure, too, but she does an admirable job of keeping it together. Midway through the party, she calls for everyone's attention and reveals a poster board on which she has written a Mad Lib about Chloe. After asking guests to supply the missing words, she reads the passage aloud, in a fit of giggles. Chloe looks delighted.

Later, Andy tells me that she has to go check on the new baby rabbits. There weren't supposed to be new bunnies right before Chloe left for college, but about a month ago, she and Serena accidentally left a male and female bunny in a cage together for a few hours. I follow Andy and a young 4-H rabbit project member over to the cages. The babies are hamster sized. Their eyes aren't open; they don't even have their trademark long bunny ears yet. We take pictures of them on our phones and coo. Chloe comes over.

"I don't know how I am going to get rid of them," she says. "Want a bunny?"

I imagine the rabbits pooping all over my living room like the turkeys did when they were little. I don't think my roommates would go for it. I tell Chloe maybe next time.

Later that week, Allison, KC, and I meet for dinner at a pizza place in Pleasant Hill. Allison's foot has healed completely, and she tells me that she ended her first semester in the independent study program with mostly A's and B's on her report card. She catches me up on Youth Fair, a 4-H show that took place a few weeks after the Contra Costa County Fair that I attended.

"Guess what?" she says. "I won small-animal masters!"

I congratulate her and she tells me all the details: the tough questions, the stiff competition from her old rivals in other clubs. It felt great to win, she says, "especially after the disaster of the first fair."

After Youth Fair, KC and Allison went on a much-needed vacation, the annual weeklong family camping trip in the Sierras. Allison spent almost every day painting ornate rocks for everyone on the trip, each with a different theme. Her older cousin loves Dr. Seuss, so she gave her a rock decorated with a Truffula tree from *The Lorax*. For the first time ever, she is actually looking forward to going back to school. She might even rack up enough credits to graduate a semester early and start some art classes at the local community college. She loves the idea that after high school, she will be able to choose what she wants to study.

"And you still have one more year of 4-H," I say. "You could win masters at next year's fair."

"Nah," says Allison. "I think if I got into masters again I'd probably let someone else do it. I want other kids to have a chance."

But Anthony, three years younger than Allison, has big plans for 4-H victory next year. When I see him in early September, he has just started high school and has made the freshman football team. I can't believe what a difference a few months have made in his appearance: he's thinned out and is now taller than his mom, and handsome. I ask him if he thinks he'll have time to practice as much with his pigs between homework and football practice. He shrugs in his laconic Anthony way and tells me he intends to win masters. He'll practice as much as it takes.

When I see Kelly at her house, she's practicing with a new lamb for an upcoming show, and her dad, Steve, is watching from a folding chair on the other side of the fence.

"Hey, Kelly," I say. "Would you call this lamb classy?"

"Yeah," she says. "I would."

Then, on a whim, I ask her if I can try cataloging the ways in which this sheep is classy.

"Go for it," she says.

I take a deep breath.

"I notice that the neck is long," I begin. "So that's a mark of growth, and suggests this lamb is going to get bigger? And I notice that it's narrower in the front and wider in back, and that's good. And I notice that all four feet are facing forward, not turned out, which is probably partially because you've set him up. Am I missing anything?"

Steve, who is laughing from his lawn chair, calls out, "You're ready to be a 4-H'er."

"I know," I say. "I'm trying!"

I often think about the observation of Raymond Lyon, the seventy-five-year-old former 4-H leader, that not much has changed in the show ring for fifty years. "You can listen to them talking, and they're talking the same stuff," he said. "It's really unusual, how it's the exact same things."

Surely that's part of what first attracted me to 4-H—the idea that a bastion of old-fashioned American farming still exists. In some ways it's true. On the surface, the shows that I saw at the fairs probably looked pretty similar to shows half a century ago—kids in whites leading animals around a show ring.

But after a few minutes, the nostalgic veneer melts away. The kids text one another from across the barn. The parents arrive in the evening looking frazzled, some of them still in their suits from the office. Sometimes they bring take-out containers from restaurants in the strip malls down the street, which were likely built on land where cattle once grazed. When they leave the barn in the evenings, many will go back to houses and apartments in the suburbs and exurbs, not farms out in the country.

When 4-H began, the organization had one goal: to bring science to farmers. Today, with seven million members around the world, its task is much more complicated. When I first learned about 4-H, I thought I had found a genuine American relic, a throwback to a simpler time. I couldn't have been more wrong.

Some of the things that I learned about 4-H culture today were profoundly worrying. I wondered whether the AgriScience curriculum benefited its corporate sponsors more than 4-H'ers. Similarly, I was concerned that the 4-H'ers I met in Ghana were not getting as much out of the hybrid maize program as DuPont was. Like Sally Pereira-Cox, Anthony's pig group leader, I didn't think it was fair that the kids who could afford expensive animals and feed stood a better chance of winning prizes than those with fewer resources. At Anthony's pig show, the kids gave vague and unstudied answers to the judge's question about the importance of farming. They seemed so focused on winning that they had lost sight of the larger point of raising animals.

But even more concerning to me were the lessons that I saw some 4-H'ers learning about food production. The AgriScience curriculum's emphasis on industrial solutions to farming problems, the widespread use of dodgy growth supplements among 4-H'ers who raise pigs, the partnership with DuPont in Ghana to promote that company's seeds—all of those things seemed like shortsighted answers to the question of how to feed a growing global population. If our future food policy experts are taught to think more about increasing the *volume* of food than about how to grow it sustainably and distribute it equitably, they'll be missing a big piece of the puzzle.

Yet even amid those troubling parts of 4-H, I discovered many bright spots. I met 4-H'ers in California's Salinas Valley who were learning, for the first time, about the crops grown on the farms all around them. I found 4-H'ers in Virginia and California who discovered a market for the animals they raised without chemicals or drugs. In Oregon, I learned about urban 4-H'ers who got to experience ranch life firsthand, and rural 4-H'ers who got a taste of city life. In Austin, I discovered city kids who were learning about agriculture by raising goats at their school. Wherever I looked, I saw evidence that 4-H has succeeded in interesting American students in science—and specifically agriculture—where our school system has largely failed. In Ghana, too, I found young people who were excited about the prospect of becoming farmers.

But I was most impressed by how 4-H enriched the lives of my core group of California livestock raisers: Allison, Anthony, Chloe, Serena, and Kelly. I consistently marveled over how smart and dedicated they were about caring for their animals, calculating proper feed ratios, keeping pens clean, and spending time with their animals every day.

Many 4-H parents and leaders talk about how the real purpose of livestock projects actually has nothing to do with the livestock, that it's all about teaching character development and leadership and responsibility. Indeed, Andy Hawkey often told me that a younger version of Chloe was worryingly timid. She wrote to me in an e-mail that another 4-H parent, one who had seen Chloe in the club for more than a decade, had recently remarked on Chloe's transformation. "She was talking about how Chloe as a little girl was so painfully shy, and couldn't maintain eye contact with anyone . . . and look at her now," Andy wrote. As I watched Chloe move seamlessly among the guests at her going-away party, I had a hard time imagining this poised young woman as a meek kid. Andy attributes her transition—at least in part—to the confidence that Chloe gained in 4-H.

And yet, I don't think that leadership and self-confidence are the only things that kids learn from 4-H. A few weeks before she left for college, Chloe let me read her admissions essay, which is about how at first, the idea of Rabbit Bowl seemed a little bit silly and trivial to her. "Why exactly does one need to know which ribs fuse to cause a pigeon breast?" she wonders. But then, she writes, something changed:

> The more I studied, the more I came to understand the idea behind the competition—to teach young rabbit breeders the information that they need to know in order to raise healthy, show-quality animals. And I loved every aspect of it. I loved the feeling as the competition drew nearer that I could handle whatever question was asked because my brain was so overflowing with rabbit knowledge; I loved the days when I could recommend a treatment to a sick rabbit or help a new member with her rabbit's breed characteristics using the facts I had learned for the competition.

I believe that the animal component of 4-H is profoundly meaningful. I witnessed my 4-H'ers develop something many kids don't: a deep sense of responsibility for other creatures on the planet—and an intimate under-standing of how those creatures affect their lives.

Most of the 4-H'ers that Raymond Lyon knew fifty years ago might have grown their own food in kitchen gardens. Today, even the 4-H'ers who live in agricultural communities—like Randy Sosa of the Salinas Valley—buy their food at the supermarket. As we have become further removed from the origins of our food, farming has undergone some of the biggest changes since 4-H was created during the early part of the last century. Genetically modi-fied crops are transforming how farmers think about yield, pests, and weeds. As climate change intensifies, more people are worried about livestock's outsized carbon footprint. In the developing world, subsistence farms are giving way to commercial operations.

The challenge for 4-H, then, is to teach critical thinking about agricul-ture—not just to kids who are lucky enough to be able to raise animals, but to everyone. In children like Lilly—the little girl who thought that chocolate milk came from brown cows—4-H has a challenge: to bridge the ever-widening divide between consumers and the people who make our food. The members of 4-H include future farmers and lawmakers, entrepreneurs and scientists, environmentalists and economists. In these kids, 4-H leaders have the chance to use their history, resources, reach, and expertise to help lay the foundations for an equitable food system in the future.

I hope they take it.

Afterword

When I try to imagine my original ideal 4-H'er now, I find that I can't do it. She has been replaced by all the actual 4-H'ers I know. Luckily for me, they're much more interesting:

As this book was going to press, **Chloe Hawkey** had transferred from Whitman College to Barnard College, where, as a sophomore, she was studying hard and exploring New York City. **Serena Hawkey**, a high school junior, was juggling AP classes and her many other activities—among which was her latest 4-H project of bringing goats to schools in some of Oakland's urban neighborhoods. After graduating from her independent study high school program, **Allison Jefferson** enrolled in classes at a local community college, where she quickly won several awards for her art. She also was awarded a 4-H prize that honors members who have done an outstanding job of teaching younger children in their clubs. **Anthony Cannon,** a high school sophomore and captain of his high school's JV football team, won a state-level 4-H award for his use of science in his swine projects. **Kelly Semonsen** was taking preveterinary classes as a freshman at her first-choice college, University of California, Davis. **Randy Sosa** of Greenfield 4-H was attending the Naval Academy Preparatory School in Newport, Rhode Island, and planning to enroll in the United States Naval Academy the following year. **Francis Baah** of 4-H Ghana was still a student at Ehiamankyene School and hoping to someday attend university to study farming.

I owe a debt of gratitude to the members of all the 4-H clubs that I spent time with, but in particular to my main subjects and their families: the Hawkeys; Allison's mother, KC Chatham; the Cannons and Sally Pereira-Cox; and the Semonsens. My agent, Judy Heiblum of Sterling Lord Literistic, helped me figure out what I was trying to say. Thank you to Kate Marshall for going to bat for my book, to Dore Brown for toiling with me in the weeds, and to everyone else at the University of California Press who worked hard to make this book better. Rafael Roy captured my hard-working 4-H'ers in his incredible photos, and Maya Dusenbery was the smartest fact checker a girl could ever ask for.

Thank you to my coworkers at *Mother Jones* for covering for me during my series of mini book leaves, and especially to my bosses, Clara Jeffery and Monika Bauerlein, for allowing me to disappear from work for weeks at a time.

Casey Miner, Tom Philpott, and Jason Mark read early drafts and gave me lots of helpful feedback and encouragement, as did my partner, Nikhil Swaminathan, and my mom, Linda Butler (even as she took umbrage at my description of my hometown as "unleafy"). My dad, Tim Butler, urged me to think more deeply about my subject matter, as he always does.

I was lucky to have a writing residency at Mesa Refuge in Point Reyes Station, California, along with grants from the Fund for Investigative Journalism, the Fund for Environmental Journalism, and the Nation Institute's Investigative Fund to support my research and writing. Parts of this book came together during weekend afternoons I spent at Oakland's and Berkeley's coffee shops, including Actual Café, Arbor Café, and Café Local 123. After my long days at the fair, I don't know what I would have done without the bottomless bread and butter, reasonably priced entrées, and soothing faux-Australian decor of the Outback Steak House.

Thank you to the women with a Y, who made me sane again over New Year's, and to Kate, Sirrah, and Amelia, who made me feel loved from thousands of miles away, as did my extended family, both Butlers and Camiels. Thank you to all my California friends who put up first with my livestock talk and then with my writing-related whining; I owe you all infinite beers. The blue ribbon for whining endurance goes to Nikhil, who made up little songs about me at my whiniest, which made me laugh instead of whine.

INTRODUCTION

1. Cary J. Trexler, Alexander J. Hess, and Kathryn N. Hayes, "Urban Elementary Students' Conceptions of Learning Goals for Agricultural Science and Technology," *Natural Sciences Education* 42 (2013): 52–55.

2. James G. Leising, Seburn L. Pense, and Matthew T. Portillo, "The Impact of Selected Agriculture in the Classroom Teachers on Student Agricultural Literacy," USDA Agriculture in the Classroom (2003), accessed January 30, 2014, http://agclassroom.org/naitc/pdf/finalreport.pdf, 1.

3. It bears mentioning that the USFRA is essentially an industry group; its affiliates include corn, soybean, dairy, and livestock alliances across the nation.

4. Axel Aubrun, Andrew Brown, and Joseph Grady, "Not While I'm Eating: How and Why Americans Don't Think about Food Systems," in *Perceptions of the U.S. Food System: What and How Americans Think about Their Food* (W. K. Kellogg Foundation, 2005), 31–54.

5. Sue A. Colbath and Douglas G. Morrish, "What Do College Freshmen Know about Agriculture? An Evaluation of Agricultural Literacy," *NACTA Journal* (September 2010): 14–17.

6. Erin Smith and Travis Park, "High School Students' Perceptions and Knowledge about Agriculture Based upon Location of the High School," *NACTA Journal* (September 2009): 17–23.

7. Shawn M. Anderson, Greg W. Thompson, and Jonathan Velez, "A Qualitative Analysis of Teachers' Conceptions of Agriculture," *2010 Western AAAE Research Conference Proceedings* (2010): 213–227.

8. "4-H around the World," National 4-H Council, accessed January 30, 2014, www.4-h.org/about/global-network/; "About 4-H," National 4-H Council, accessed January 30, 2014, www.4-h.org/about/; Rama B. Radhakrishna and Megan Sinasky, "4-H Experiences Contributing to Leadership and Personal Development of 4-H Alumni," *Journal of Extension* 43, no. 6 (2005), accessed January 30, 2014, www.joe.org/joe/2005december/rb2.php.

9. "4-H Distinguished Alumni," National Association of Extension 4-H Agents, accessed January 30, 2014, www.nae4ha.com/4-h-distinguished-alumni.

10. The demographic figures in this paragraph are from "4-H Reports," USDA Research, Education, and Economics Information System, accessed January 30, 2014, www.reeis.usda.gov/portal/page?_pageid=193,899783&_dad=portal&_schema=PORTAL&smi_id=31.

11. "California 4-H Projects," University of California 4-H Youth Development Program, accessed January 30, 2014, http://4h.ucanr.edu/files/14270.pdf.

12. "National 4-H Council, Sponsor: Opportunities and Benefits," National 4-H Council, accessed January 30, 2014, www.4-h.org/get-involved/sponsor/corporate-foundation-benefits-opportunities/.

13. "World Trade in Agricultural Products, 2003," World Trade Organization, accessed January 30, 2014, www.wto.org/english/tratop_e/agric_e/negs_bkgrnd19_data_e.htm.

ONE. "I WANTED TO BE A COWGIRL"

1. Marilyn Wessel and Thomas Wessel, *4-H: An American Idea, 1900–1980* (Chevy Chase, MD: National 4-H Council, 1982).

2. Ibid., 1.

3. Ibid., 5.

4. Ibid., 4–5.

5. Olly Jasper Kern, *Among Country Schools* (Boston: Ginn and Company, 1906), 42, 87, 158, 160, 181.

6. *Among Country Schools* is full of Kern's idiosyncratic—and often forward-thinking—ideas about how to improve rural education. An entire chapter is

devoted to the importance of trees in the schoolyard, complete with photos of happy children frolicking around shrubbery and, for comparison, stern students huddled together on their "Treeless School Grounds." He believed strongly that gardening was good for kids and proposed that "instead of cities building larger jails and pointing with pride to such structures as the solution to the bad-boy problem, let more money be spent in farm schools, where the boy can get away from the slum back to the brown earth" (84).

7. Wessel and Wessel, *4-H*, 13. In her book *Every Farm a Factory: The Industrial Ideal in American Agriculture* (New Haven: Yale University Press, 2003), historian Deborah Fitzgerald argues that the government's campaign to modernize farmers was productive—but also infantilizing. At that time, farmers were popularly perceived as hayseeds, and lazy ones at that. "Even Henry Ford, the 'farmer's friend,' felt that most farmers only worked hard about thirty days a year, spending the rest of the time just fooling around on their farms," writes Fitzgerald (50). The new university Extension service, many hoped, would coax farmers out of what they saw as backward ways, and 4-H was part of this effort. Indeed, "agents created boy's corn clubs and girl's canning clubs in an effort to persuade or embarrass parents to adopt more modern practices in the fields and at home" (53).

8. Wessel and Wessel, *4-H*, 16.

9. Ibid., 17.

10. Ibid., 19.

11. Ibid., 24.

12. Steven M. Worker, "History of Science Education in the 4-H Youth Development Program," Monographs (Winter 2012), University of California 4-H Center for Youth Development, accessed April 28, 2014, http://4h.ucanr.edu/files/135384.pdf.

13. Wessel and Wessel, *4-H*, 25.

14. Members could identify as more than one race.

TWO. LEARNING BY DOING

1. This information about the history of agriculture in California's East Bay comes from an interview conducted over the phone with Diane Curry, curator and archivist of the Hayward Area Historical Society, September 20, 2013.

THREE. "I DO SHEEP THE WAY OTHER KIDS DO SOCCER"

1. Christine Souza, "Leaders in the World's Salad Bowl," *California Country Magazine* (California Farm Bureau), July–August 2009, accessed January 30, 2014, www.californiacountry.org/features/article.aspx?arID=563.

2. "Broccoli, California Grown," See California, accessed January 30, 2014, www.seecalifornia.com/farms/broccoli.html.

3. Carmen V. Harris, "States' Rights, Federal Bureaucrats, and Segregated 4-H Camps in the United States, 1927-1969," *The Journal of African American History*, 93, no. 3 (Summer 2008): 362–388.

FOUR. BRINGING UP BABY

1. Martin H. Smith and Cheryl L. Meehan, *Exploring Beef Health and Husbandry* (Chevy Chase, MD: National 4-H Council, 2013), 9.

FIVE. THE BIG BUSINESS OF 4-H

1. Marilyn Wessel and Thomas Wessel, *4-H: An American Idea, 1900–1980* (Chevy Chase, MD: National 4-H Council, 1982), 34.

2. Ibid., 37.

3. Ellen E. Moberg, "Sociability Lane," *National 4-H Club News*, March–April 1939, 10.

4. Wessel and Wessel, *4-H*, 51.

5. "History of the National 4-H Awards Program," 4-H History Preservation Program, accessed January 30, 2014, http://4-hhistorypreservation.com/History/National_Recognition/.

6. Ibid.

7. "Guy Noble," Iowa 4-H Foundation, accessed April 28, 2014, www.iowa4hfoundation.org/index.cfm/31887/2050/guy_noble.

8. "Donors Conference Focuses on Campaign for 4-H," *National 4-H Council Quarterly*, Fall 1983, 1–2.

9. "1988 Private Support for 4-H Programs," *National 4-H Council Quarterly*, Spring 1988, 15–16.

10. "Welcome to Project Butterfly Wings," Florida Museum of Natural History, accessed January 30, 2014, www.flmnh.ufl.edu/wings/index.asp.

11. Derek Riley and Alisha Butler, "Priming the Pipeline: Lessons from Promising 4-H Science Programs" (a report prepared by Policy Studies Associates for National 4-H Council, July 2012), accessed April 28, 2014, www.ca4h .org/files/154449.pdf, B-3, B-13.

12. "4-H On the Wild Side Program," University of California Cooperative Extension: Sacramento County, accessed January 30, 2014, http://cesacramento .ucanr.edu/4H/On_The_Wild_Side_Program/.

13. Annual reports from the National 4-H Council, 2007 and 2012, accessed April 28, 2014, www.4-h.org/about/annual-report/.

14. Monica Mielke and Alisha Butler, "4-H Science Initiative: Youth Engagement, Attitudes, and Knowledge Study" (a report prepared by Policy Studies Associates for National 4-H Council, March 2013), accessed April 28, 2014, www.4-h.org/About-4-H/Research/PSA-YEAK-Year-3-Report.dwn.

15. Richard M. Lerner, Jacqueline V. Lerner, and colleagues, "The Positive Development of Youth: Comprehensive Findings from the 4-H Study of Positive Youth Development" (a report prepared by Tufts University's Institute for Applied Research in Youth Development for National 4-H Council, 2013), accessed April 28, 2014, www.4-h.org/About-4-H/Research/PYD-Wave-9-2013.dwn.

16. Riley and Butler, "Priming the Pipeline," 11.

17. *4-H Fund Development Toolkit*, National 4-H Council and The Osborne Group, Inc., accessed January 31, 2014, www.4-h.org/resource-library /professional-development-learning/fund-development-toolkit-members/.

18. "Creating the Case for 4-H Science," *4-H Fund Development Toolkit*, National 4-H Council and The Osborne Group, Inc., accessed January 31, 2014, www.4-h.org/Professional-Development/Fund-Development-Toolkit/major-gifts/maximizing-corporate-giving/Creating-the-Case-for-4-H-Science.dwn.

19. Fawn Roark, "4-H Program Receives Afterschool Grant," *The Mountain Times*, November 9, 2006, accessed January 31, 2014, http://archives .mountaintimes.com/mtweekly/2006/1109/4h.php3.

20. "AgriScience," National 4-H Council, accessed January 31, 2014, www .4-h.org/resource-library/curriculum/agriscience/.

21. "Food and Science Technology: What Makes NesQuik™ Quick?" (curriculum for National 4-H Council AgriScience program), National 4-H Council,

accessed January 31, 2014, www.4-h.org/WorkArea/DownloadAsset .aspx?id=7258&libID=7253.

22. "Face the Fat: Engineering a Better Oil" (curriculum for National 4-H Council AgriScience program), National 4-H Council, accessed January 31, 2014, www.4-h.org/WorkArea/DownloadAsset.aspx?id=7256&libID=7251.

23. "Agriculture at Work: Bioplastic" (curriculum for National 4-H Council AgriScience program), National 4-H Council, accessed January 31, 2014, www.4-h.org/WorkArea/DownloadAsset.aspx?id=7241&libID=7236.

24. "Teens Teaching Youth AgriScience/Biotechnology: Results, Recommendations, and Promising Practices" (National 4-H Council report, 2013), accessed April 28, 2014, www.4-h.org/workarea/downloadasset.aspx?id= 64730, 5.

25. Charles M. Benbrook, "Impacts of Genetically Engineered Crops on Pesticide Use in the U.S.—the First Sixteen Years," *Environmental Sciences Europe* 24 (2012), accessed April 28, 2014, www.enveurope.com/content/24/1/24.

26. "DuPont and National 4-H Council Announce Youth Development Initiative for Rural Africa" (National 4-H Council press release, October 11, 2011), National 4-H Council, accessed January 31, 2014, www.4-h.org/About-4-H /Media/Press-Releases/DuPont-and-National-4-H-Council-Announce-Youth-Development-Initiative-for-Rural-Africa/.

SIX. "WE ARE PRAYING THAT DUPONT WILL CONTINUE TO PROVIDE FOR US"

1. Gabriel N. Rosenberg, *Breeding the Future: 4-H and the Roots of the Modern Rural World* (Philadelphia: University of Pennsylvania Press, forthcoming).

2. Law and his colleagues also promoted a political agenda, giving the leaders "special training in 'democratic' leadership that would 'guide them away from being autocratic,'" according to Rosenberg's *Breeding the Future.* "Law and many of the funders also explicitly considered 4-H clubs as effective anti-communist organizations," Rosenberg wrote to me in an e-mail.

3. "Cultivating Learning with School Gardens" (curriculum for 4-H Ghana's Enterprise Garden program, July 2011), National 4-H Council, 18–19.

4. Pedro A. Sanchez, Glenn L. Denning, and Generose Nziguheba, "The African Green Revolution Moves Forward," *Food Security* 1 (2009): 37–44.

1. "Specialty Products for the Look of a Champion," Sunglo Feeds, accessed February 1, 2014, http://sunglofeeds.com/products/.

2. "Show Pig Supplements," Essential Feeds, accessed February 1, 2014, www.essentialshowfeeds.com/essential_show_pig_supplements.

3. Helena Bottemiller, "Dispute over Drug in Feed Limiting US Meat Exports," Food and Environment Reporting Network, January 25, 2012, accessed April 28, 2014, http://thefern.org/2012/01/dispute-over-drug-in-feed-limiting-u-s-meat-exports/.

4. *Elanco Animal Health: Food and Companionship Enriching Life* brochure (Greenfield, IN: Elanco Animal Health, 2010), accessed February 1, 2014, www.elanco.com/pdfs/2010_10869_elanco-corporate-brochure-ai-10988.pdf, 8.

5. Gloria J. Dunnavan, director, Division of Compliance, Center for Veterinary Medicine, FDA, to Patrick C. James, president, Elanco Animal Health, September 12, 2002, accessed February 1, 2014, www.fda.gov/ICECI/EnforcementActions/WarningLetters/2002/ucm145110.htm.

6. J. N. Marchant-Forde, D. C. Lay, E. A. Pajor, B. T. Richert, and A. P. Schinkel, "The Effects of Ractopamine on the Behavior and Psychology of Finishing Pigs," *Journal of Animal Science* 81 (2003): 416–422.

7. R. Poletto, M. H. Rostagno, B. T. Richert, and J. N. Marchant-Forde, "Effects of a 'Step-Up' Ractopamine Feeding Program, Sex and Social Rank on Growth Performance, Hoof Lesions and Enterobacteriaceae Shedding in Finishing Pigs," *Journal of Animal Science* 87 (2009): 304–313; R. Poletto, H. W. Cheng, R. L. Meisel, J. P. Garner, B. T. Richert, and J. N. Marchant-Forde, "Aggressiveness and Brain Amine Concentration in Dominant and Subordinate Finishing Pigs Fed the B-Adrenoreceptor Agonist Ractopamine," *Journal of Animal Science* 88 (2010): 3107–3120.

8. Helena Bottemiller, "Ractopamine and Pigs: Looking at the Numbers," Food and Environment Reporting Network, February 23, 2012, accessed February 1, 2014, http://thefern.org/2012/02/ractopamine-and-pigs-looking-at-the-numbers/.

9. "FDA/CVM-ADE Reports-CVM Response," Scribd, via Food and Environment Reporting Network, April 18, 2011, accessed February 1, 2014, www.scribd.com/doc/82609189/ADE-Pig-Samples.

10. Ibid.

11. "Strategies for Feeding Ultra Modern Showpigs," Honor Show Chow, accessed February 1, 2014, http://honorshowchow.com/feedingstrategyarticles /HSC%20Strategies%20For%20Feeding%20Modern%20Showpigs.pdf.

12. David A. Kessler, "Antibiotics and the Meat We Eat," *New York Times*, March 27, 2013, accessed February 1, 2014, www.nytimes.com/2013/03/28 /opinion/antibiotics-and-the-meat-we-eat.html.

13. Jay P. Graham, John J. Boland, and Ellen Silbergeld, "Growth Promoting Antibiotics in Food Animal Production: An Economic Analysis," *Public Health Reports* 122 (2007): 79–87.

NINE. THE CONTRA COSTA COUNTY FAIR

1. "Foreclosure Rates for 94531," RealtyTrac, accessed February 9, 2014, www .realtytrac.com/statsandtrends/foreclosuretrends/ca/contra-costa-county/antioch /94531; Chris Schmidt and Jake Wegmann, "Diversity Didn't Cause the Foreclosure Crisis," SPUR, June 3, 2012, accessed February 9, 2014, www.spur.org/publications /article/2012-06-03/diversity-didn-t-cause-foreclosure-crisis.

2. Robin Li, interview with Raymond G. Lyon, *Taking the University to the People: University of California Cooperative Extension,* 2009, Regional Oral History Office, The Bancroft Library, University of California, Berkeley, accessed February 9, 2014, http://digitalassets.lib.berkeley.edu/roho/ucb/text/lyon_ raymond.pdf.

3. "The Incredible Pig: Swine Youth Activity Guide, Grades 3–5" (curriculum for National 4-H pig projects, 2004), National 4-H Council, 11.

ELEVEN. "IT'S IN MY BLOOD"

1. "Going Whole Hog" (curriculum for National 4-H pig projects, 2004), National 4-H Council.

2. "Nationwide Surveys Reveal Disconnect between Americans and Their Food," U.S. Farmers and Ranchers Alliance, September 20, 2011, accessed February 2, 2014, www.fooddialogues.com/2011/09/22/nationwide-surveys-reveal-disconnect-between-americans-and-their-food.

3. Shawn M. Anderson, Greg W. Thompson, and Jonathan Velez, "A Qualitative Analysis of Teachers' Conceptions of Agriculture," *2010 Western AAAE Research Conference Proceedings* (2010): 213–227.

4. "4-H Reports," USDA Research, Education, and Economics Information System, accessed January 30, 2014, www.reeis.usda.gov/portal/page?_pageid=193,899783&_dad=portal&_schema=PORTAL&smi_id=31.

5. *4-H Urban-Rural Exchange for a Sustainable Future*, 4-H Programs of Distinction Online Database, 4-H National Headquarters, accessed April 28, 2014, www.csrees.usda.gov/nea/family/res/pdfs/4_h_database/4-H_Urban-Rural_Exchange_2009.pdf.

6. "Crossing the Urban-Rural Divide," *Oregon Field Guide*, season 21, episode 1, Oregon Public Broadcasting, December 20, 2012, accessed April 28, 2014, http://watch.opb.org/video/2315334787/.

7. "What Do You Know about Animal Science?," test by Texas A&M AgriLife Extension and 4-H Capital, provided by Lydia Domaruk.

8. The answer to the first question is b (16%), and the answer to the second question is d (wool).

9. "Science of Agriculture Programs," a summary of 4-H Capital's evaluations of its programs, provided by Lydia Domaruk.